D1145872

WITHDRAWN
IOWA STATE UNIVERSITY
LIBRARY

WITHDRAWN
IOWA STATE UNIVERSITY
LIBRARY

Imaging Radar
for
Resources Surveys

WITHDRAWN
IOWA STATE UNIVERSITY
LIBRARY

REMOTE SENSING APPLICATIONS

Series Editors

E. C. Barrett
Director, Remote Sensing Unit
University of Bristol

L. F. Curtis
Exmoor National Park Officer
and
Honorary Senior Research Fellow,
University of Bristol

In the same series:

Remote Sensing of Ice and Snow
Dorothy K. Hall and Jaroslav Martinec

Imaging Radar
for
Resources Surveys

J. W. TREVETT

London   New York
CHAPMAN AND HALL

First published in 1986 by
Chapman and Hall Ltd
11 New Fetter Lane, London EC4P 4EE
Published in the USA by
Chapman and Hall
29 West 35th Street, New York, NY 10001

©1986 J. W. Trevett

Printed in Great Britain at the
University Press, Cambridge

ISBN 0 412 25520 0

All rights reserved. No part of this book may be reprinted, or reproduced or
utilized in any form or by any electronic, mechanical or other means, now known
or hereafter invented, including photocopying and recording, or in any
information storage and retrieval system, without permission in writing from the
publisher.

British Library Cataloguing in Publication Data

Trevett, J. W.
    Imaging radar for resources surveys.
    1. Natural resources—Remote sensing   2. Radar
    I. Title
    333.7'028   HC79.R3/
    ISBN 0–412–25520–0

Library of Congress Cataloging in Publication Data

Trevett, J. W., 1927–
    Imaging radar for resources surveys.

    Bibliography: p.
    Includes index.
    1. Remote sensing.   2. Radar.   I. Title
    G70.4.T74   1986   621.3678   85–29899
    ISBN 0–412–25520–0

To K., S. K. and L. A. E. for persistence and faith.

Contents

Preface

The use of air photographs as an aid to understanding and mapping natural resources has long been an established technique. The advent of satellite imagery was, and indeed by many still is, regarded as a very high altitude air photograph, but with the introduction of digital techniques the full analysis of imagery has become very sophisticated.

Radar imagery presents the resource scientist with a new imaging technique that has to be understood and used, a technique which, although in many respects still in its infancy, has considerable applications potential for resources studies.

Remote sensing now forms an element in study courses in the earth sciences in many major universities and a number of universities offer specialist post-graduate courses in remote sensing. Nevertheless there are a large number of earth scientists already working with imagery who have progressed from the air photograph base to satellite imagery. Such scientists may find themselves confronted with microwave or radar imagery or wish to use the imagery for surveys and find themselves hindered by a lack of understanding of the differences between radar imagery and optical imagery. Unfortunately reference to much of the literature will not be of very great help, many excellent text books on the theory and interaction of microwaves, on instrument design and construction and on the research carried out on specific target types exist, most of these are however written for specialists who are usually physicists not earth scientists.

This book is therefore a modest attempt to try to provide the earth scientist with some background to microwave technology, to demonstrate by reference to studies already carried out the applications to some key disciplines, to give an insight into future developments and application and hopefully to awaken interest in the subject such that they will then be encouraged to read some of the more specialist literature in order to study the subject further.

The earlier chapters have been designed to alert the interpreter to the methods of data acquisition, the differences from optical imagery and the problems involved in processing data. To the physicists the descriptions may appear to be brief and even to be treated lightly, but it is not the intention of this book to compete with the more detailed and established text books on the subject written by far more eminent researchers than this author. To assist those who would like to pursue the subject further an extensive bibliography has been included in the volume. In this context the early chapters are intentionally largely free of references, this is partly a matter of style – the continuous introduction of references is considered to be intrusive on the flow of reading – and partly because a conclusion may be a consensus of several references and attributing the component parts is extremely difficult.

Later in the volume where case histories are used to describe application the references are more meaningful and have been included in the text.

Every effort has been made to include the most up-to-date advances or techniques on the subject, nevertheless, when dealing with a rapidly developing technology, it is inevitable that some new development will have been introduced which makes some of the techniques described obsolete before the printing ink is dry. The fundamental theories should remain static and the project descriptions will still be relevant to the processes of interpretation.

It is hoped that this book will introduce more earth scientists to the potential of radar imagery, and that it will encourage them to advocate its use for surveys with which they are connected, to study the subject further and to make their own contributions to the research.

In preparing this book the author is indebted to many people and organizations. A number of companies and research establishments gave freely of their material and hopefully adequate acknowledgement of this is made throughout the book. Without much of this help and material the subject would have been less well illustrated and described. Special acknowledgement must be made to Hunting Technical Services Limited for their support, through the Managing Director, Don Chambers, in the production of this book; the report section undertook most of the draft typing and the high standard of the diagrams is due to the work of the cartographic section, my thanks therefore to the individuals concerned: Nigel Schofield, Dora Godfrey, Cheryl Coffey, Pat Bridge and Elaine Quin, and to Alan Lonslow and Alison Thwaites for the drawings. Grateful acknowledgement is also made to Hunting Technical Services and the sister company Hunting Geology and Geophysics for permission to reproduce much valuable material from work undertaken by them. I would also like to express my appreciation to Peter Churchill who was persuaded to cast a critical eye over the drafts and made many useful comments. Finally there is my long suffering family who have helped with some of the typing, provided considerable encouragement and have silently endured a house filled with reference material.

1

An introduction to imaging radar

1.1 INTRODUCTION

The application of conventional air photographs to the mapping of the earth's resources has been an established system for a considerable number of years. Initially the main impetus for air photo-interpretation came from the topographic mapping community who were looking for a quick and reliable alternative to laborious field methods of map production.

In the period following World War II the emphasis changed to the need to develop natural resources and the use of the photographic image for this purpose has developed out of this increase in activity. For many resources studies the conventional panchromatic photograph did not provide sufficient data, with the result that colour photography, false colour infra-red photography and multi-spectral imagery have all been developed until we now have the present series of high resolution multi-spectral optical imaging satellites. All these developments have increased the ability to map the world's resources and to plan their development but optical devices are dependent upon the reflectivity of surfaces to the sun's rays and are of no value when there is cloud cover.

The research work of Taylor, Young, Page, Watson Watt and Guthrie in the 1920s and 1930s prepared the way for the rapid development of radar's military application during World War II and subsequently. During the early use of radar it was observed that, other than the specific military target, objects such as buildings, landforms and forests, produced a return or echo back to the receiver. This return was called ground clutter by the radar engineer and attempts were made to filter it out of the system. As radar was declassified and became more widely used, scientists realised that ground clutter patterns provided a 'new look' at the earth's surface. What is more, radar had the advantages of not relying upon the sun's illumination to produce an image and of being unaffected by haze and most clouds.

Airborne imaging radars were subsequently developed and a series of development contracts issued by the US military authorities resulted in the construction of three different airborne imaging systems which effectively demonstrated the potential of radar as an alternative imaging system.

During the past decade imaging radars have been improved and have been operated in orbiting satellites and in the space shuttle. At the same time earth resources scientists have studied and used the imagery and it is now recognized as a viable alternative imaging system of particular benefit in cloud covered areas.

Photographs produce images which are comparable to the view we normally see, since our eyes and the camera are both recording reflected sunlight. Thus the interpreter has little difficulty in recognizing the observed image, although training may well be needed in the deductions made from the observed data. What is more, the science of photography is well advanced and the interpreter does not have to have a deep understanding of photographic processing in order to make an interpretation of an image.

Radar, on the other hand, is not recording the normally observed view, although the image may present many elements that are recognizable by virtue of their size or shape. The waveband frequency used in the radar, the strength of the signal, the receiving antenna, the processing used and the object viewed can all affect the eventual representation on the image. The interpreter must as a result understand something of the nature of radar imaging systems and processing in order to carry out a meaningful interpretation. It is for this reason that this book commences with a description of basic radar principles. The description is not intended to be an in depth study of radar, for this the reader is referred to the bibliography where the main text books are listed, the intention is to give the average resource interpreter some background to the data base they will be using.

1.2   THE ELECTROMAGNETIC SPECTRUM

Energy is propagated by electromagnetic radiation which is transmitted directly through free space or indirectly either by reflection or re-radiation to a sensor. Different forms of energy radiate at different wavelengths and the types are usually defined by their wavelengths.

Figure 1.1 shows the standard electromagnetic spectrum which is given against a measure of wavelengths. Although the electromagnetic spectrum is very wide and is shown ranging from gamma rays to very high frequency

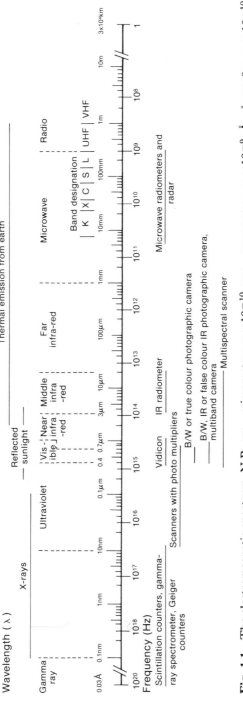

Fig. 1.1 The electromagnetic spectrum. N.B. μm – micrometres $m \times 10^{-10}$; nm – nanometres, $m \times 10^{-9}$; Å – ångström, $m \times 10^{-10}$.

(VHF) radio waves, it is only feasible to use specific segments of the spectrum for imaging systems. These are usually seen as three segments:

(1) The visible or reflected sunlight range which is the region imaged by cameras and multi-spectral scanner systems and which includes a far-red or near infra-red range;

(2) The thermal infra-red range which includes some thermal sunlight reflection but also includes natural thermal emissions from the earth, as in volcanoes;

(3) The microwaves or radars.

The transparency and effect of the atmosphere on the transmission of electromagnetic energy restricts the regions which are used to those defined above, these are the so called atmospheric windows.

Within the visible spectrum, light, as we observe it, can be divided into different wavelengths which we observe as different colours, thus the eye which receives energy in different wavelengths and transmits them to the brain as different colours is a form of multi-spectral scanner.

Microwaves cover a more extensive range of the electromagnetic spectrum and are usually divided into designated spectral bands, K, X, C, S and L as shown; the human brain however is not capable of registering these through the eye, nor are they related to colours. There is no particular significance in the notations chosen for the different radar bands, the random choice of letters originates from a World War II decision intended to confuse the opposition, the notation has been retained since.

Although microwaves occur naturally they are not normally used to provide an image, in the same way that a high energy source such as the sun is required to illuminate a surface for optical imaging, so it is necessary to generate a high energy microwave source to illuminate targets for radar imaging. What is eventually measured is the returned or backscattered energy from a target. Microwave sensors are therefore referred to as active microwave sensors when they provide their own energy source and as passive microwave systems when they measure radiation from other sources. In active microwave sensors it is normal to combine the two operations so that one platform will both generate the microwave energy and receive the backscattered signal.

1.3 RADAR SYSTEMS

The acronym RADAR is derived from 'radio detection and ranging' and it is therefore a device which is capable of detecting an object (the target), indicating its distance (range) and its position (direction). The term radar is commonly used for all active microwave systems although they may be non-ranging devices.

It has already been stated that an active system uses its own energy source, this is in the form of a micro radio wave which can retain its form and strength when travelling through the atmosphere, i.e. where the atmosphere does not have any alternative influences. The wavelengths used can penetrate haze, cloud, smoke, heat shimmer and even light rain and still retain their waveforms.

The normal configuration of a radar imaging device can be seen in Fig. 1.2 and consists of a transmitter, a modulator, a receiver and a processor usually with a display unit. The systems are normally designed so that only one antenna is used for both transmitting and receiving.

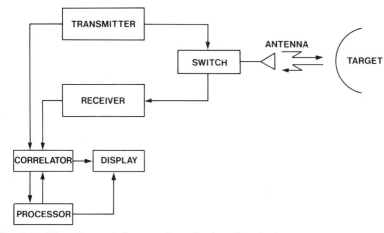

Fig. 1.2   The schematic layout of a radar imaging device.

The energy is transmitted as short but powerful bursts, known as pulses, which are only microseconds in length. These pulsed signals travelling at $300\,000$ km s^{-1} strike the targets in the field of view where some of the energy is absorbed, some is reflected away and some refracted or backscattered to the receiver. The returning signal will be weaker than the transmitted signal and the transmitter has to be turned off to minimize interference and enable the returning signal to be detected, the pulse radars are therefore continually being switched off and on in phase with the pulses.

In the receiver the strength of the returning signal is measured, this in turn will provide the grey scale tone of the image, at the same time the return time of the signal is calculated and this determines the distance of the reflecting object from the receiver. These two functions of signal strength and delay time constitute the main elements of a radar image. The signal return data are subsequently recorded either as an optical record or a digital signal or both.

With optical systems using the sun for illumination the transmitter (the sun) and the receiver (the camera) are in two different locations and under normal circumstances the imagery is obtained at a time when the sun is producing a convenient shadow to highlight topographic relief. The imaging device can be given a vertical view in these circumstances and the imagery is thus devoid of unnecessary geometrical distortions and nearer to a correct orthogonal view. If radar was to be used in a vertical mode, because the energy would be transmitted downwards, there would be very little relief effect derived from shadows, at the same time energy would be directly reflected back to the antenna and not backscattered.

Radar transmission is therefore at an oblique angle to the platform, this generates artificial shadows and ensures backscattering. The angle of view known as the depression angle is varied for different requirements and is a subject for much research in order to study the effect on the image for different targets. It is this oblique view of the radar which led to the description side looking airborne radar (SLAR) which is used to describe all such systems. SLAR is an airborne sensor for displaying the radar backscatter characteristics of the earth's surface in the form of a strip map or picture of a selected area (Moore, Manual of Remote Sensing, 1983).

Imaging radars fall into two different categories: real aperture, with which the term SLAR is often associated, although RAR is now more commonly used, and synthetic aperture radars known by the acronym SAR. In the following sections the term SLAR will be used to cover common features and the terms RAR or SAR used for the individual systems as appropriate.

1.3.1 RAR

Real aperture radars use a narrow bandwidth of energy to observe the targets and are very similar to the normal form of pulse radars used for example to observe and monitor shipping or aircraft. In these latter systems the antenna sends out a narrow beam pulse which reflects from a target back to the antenna to be subsequently displayed upon a cathode ray tube screen. The antenna is rotated in order to provide coverage over an area and the display shows shipping and a coastline, for example, within the observed area.

Rotating antenna systems are used on aircraft, most large aircraft contain such systems in order to provide the pilot with advance data on storms. There is however a limit to the size of rotating antenna that can be mounted on an aircraft and this restricts the degree of definition that can be obtained. A fine angular resolution can be obtained only with a large aperture with respect to the wavelength of received radiation.

High resolution RARs are obtained by mounting a long fixed antenna under the belly of the aircraft with the forward movement of the aircraft replacing the rotational beam. The antenna can be as long as 5 m but the ultimate effective length of the antenna will be conditioned by the size of the aircraft to be used and the effect of the antenna on the flight performance of the aircraft. In some of the longer wavelengths used where the antennna aperture length is equivalent to 100 times the wavelength, the ultimate antenna could be too large for any aircraft.

The target responses are continuously displayed by a spot scanning across a cathode ray tube, the light intensity of the spot being modulated in accordance with the strength of the returning signal. By passing film across the tube at a speed related to the forward motion of the aircraft, a continual record of the returned signals can be made for subsequent study. Some systems employ a more conventional television display system to provide a real time imaging display and the same technology can be used to provide a video recorded image for subsequent study.

The latter systems are frequently used for low cost radar imaging systems designed for specific surveillance, as in oil pollution monitoring. In these instances the operator is only interested in identifying the occurrence of oil spillage and is not so concerned with an accurate permanent record of the sea surface.

Systems have also been designed that transmit the signal directly to a ground station, this may be required to monitor the survey from more than one imaging aircraft for example, or to enable a patrol vessel to proceed to the pollution area and to identify its own progress relative to the target such as an oil spillage, which may not otherwise be visible from the ship. (The ship will itself appear as a target within the observed area).

RARs use a narrow swath to observe the target area but, since they are side looking, the width of the swath will increase across the area. This in turn means that the resolution of the system will decrease with distance from the platform. It is possible to contain this within acceptable tolerances when using an airborne system but it does impose a constraint which becomes unacceptable for potential satellite borne systems.

There is an obvious advantage in any real time system, but the displayed image range is reliant upon the dynamic range of the display and, more particularly, on film emulsions. There is a limit to the number of grey levels that can be recorded onto film and this means that the full potential of a system may not be used. In recent years digital scan converters have been developed (e.g. University of Kansas by Moore et al.) which enable the data to be recorded in a digital mode for subsequent computer analysis.

Unfortunately, to achieve an acceptable resolution and dynamic range, RAR requires both a long antenna and a high power output, hence RARs are sometimes referred to as 'brute force' systems. The power specifications can usually be met with airborne systems, indeed the Motorola system is an example of a viable RAR, but they are difficult to satisfy for satellite or other space craft systems where there are constraints on available power.

1.3.2 SAR

It has been shown that RARs have severe constraints imposed upon them caused by both the size of antenna that can be accommodated, which in turn affects wavelengths used, and by the ability to maintain resolution of over-long operating ranges.

Synthetic aperture radar (also known as coherent SLAR) is a means of overcoming these problems. The SAR uses a relatively short antenna but synthesizes a much larger antenna by taking advantage of the forward motion of the platform on which it is housed. The concept of the synthetic aperture radar was originated by Wiley and was originally referred to as the Doppler beam sharpener.

Although the SAR antenna is shorter than the RAR it nevertheless has a wider field of view or beam swath. A target will therefore enter into the beam and as the platform moves past the target, will remain in the field of view for a period of time before passing out of the beam. The SAR like the RAR is side looking and the beam will increase in width across the observed area with distance from the platform. The result is that a target at a distance from the platform will remain in the field of view longer than a target closer to the platform.

It was indicated with RARs that their resolution potential was limited by the size of the antenna, as resolution is proportional to the length of the antenna but inversely proportional to the range. The SAR effectively increases the antenna length with range, the target remains in view longer the further it is away, with the result that the resolution of a SAR remains effectively the same over all ranges, an attractive element when looking for a space imaging system.

The effect of the synthetic aperture antenna with a target can be seen in Fig. 1.3; in this instance the target is shown being recorded onto film. SAR recording can be onto film or onto digital tape. The figure shows how a long antenna can be synthesized by a short antenna by taking advantage of the airplane's motion. As the airplane flies along a straight line a short real antenna mounted on its belly sends out a series of pulses at regular intervals. Each pulse consists of a train of coherent microwaves (*dark curves*). Although the length of the pulses determines the resolution across the track, it is the wavelength of the microwave radiation that determines the resolution along the track. As an object (*black dot*) enters the antenna's beam (*left*) it reflects the portion of the pulses it receives back toward the antenna. At some points in the aircraft's path the object will be an integral number of microwave wavelengths away; between those points it will not be. In the illustration the object is first 11 wavelengths away (a), then 10 (b), then 9 (c), then $8\frac{1}{2}$ (d), at which point the object is at right angles to the antenna. From then on the airplane is increasing its distance from the object (e). The

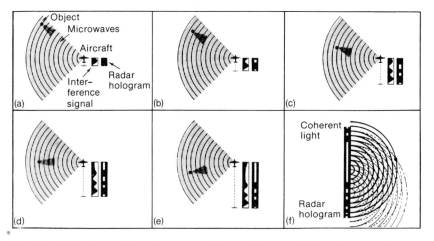

Fig. 1.3 The interaction of a target with an airborne SAR on its recording as a hologram on film. (Courtesy of Goodyear Aeroservices.)

antenna receives the series of reflected waves (*curves in light colour*) and electronically combines them with a train of reference wavelengths (not shown), causing the two series of oscillations to interfere. The interference signal emerges as a voltage, which controls the brightness of a spot scanning across the screen of a cathode-ray tube. At the times that a returned pulse coincides with a reference pulse the interference is constructive; the voltage will be high and the moving spot will be bright. At the times that the phase of the returned wavelength does not coincide with the phase of the reference frequency the interference is destructive; the voltage will be low and the moving spot will be dim. The moving spot thus traces out a series of light and dark dashes of unequal length, which are recorded on a strip of data film moving at a velocity proportional to the velocity of the airplane. The series of opaque and transparent dashes on the film is actually a one-dimensional interference pattern; the film on which they are recorded is a radar hologram. When the developed hologram is illuminated by a source of coherent light (f), each transparent dash will act as a separate source of coherent light. Behind the hologram there will be a single point where the resulting light waves all constructively interfere. There the 11th wavelength of light (*thin curves*) from the transparent dash created by the 11th microwave will meet the 10th wavelength of light (*broken curves*) from the transparent dash created by the 10th microwave, and both will meet the 9th wavelength of light (*thick curves*) from the transparent dash created by the 9th microwave. At that one point light from the entire length of the interference pattern is focused to form a miniature image of the original object. Although this produces a photographic image, the resulting data would be virtually incomprehensible to an interpreter without further processing. An essential element of SAR therefore is the need for subsequent complex data processing whether of the optical image or the digital record. This processing is considered in more detail in Chapter 4. The use of the synthetic aperture can be shown to be capable of considerably improving image resolution since the theoretically achievable resolution is independent of the operating range and the resolution can be further improved by reducing the physical dimensions of the antenna.

These advantages would appear to make SAR the ideal radar imaging system but as with all apparent major benefits there has to be some trade off or potential disadvantages. SAR is no exception, film recording is limited by the dynamic range of photographic materials and there is a point of maximum data recording in terms of grey scale and resolution after which any refinements in the system cannot be matched by film recording media. Digital recording offers the ability to record a greater dynamic range but the volume of digital data greatly increases, only by the use of high density digital tapes (HDDTs) can the data be recorded and even then there are limitations. Subsequently the volume of digital data introduces further problems in the amount of digital processing required to produce processed computer compatible tapes (CCTs) which can be used to produce photographic images or analysed using digital image analysers.

As the system is refined to improve the resolution and the antenna length is decreased, so the power requirement is increased which in turn leads to a greater signal/noise ratio; there has to be a point therefore where this ratio can just be handled and greater resolution achieved.

1.3.3 Non-imaging radars

Both RAR and SAR are imaging radar systems, and in practical terms these are the systems which will concern most resources scientists. There are however radars which do not provide images, the non-imaging systems; the radar altimeter can be seen as one example of such a system. The radar altimeter is used to measure the distance from the platform, usually an aircraft or satellite, to the target, in this case the earth's surface, and to record the data as a digital record or a continuous profile. Radar altimeters are of particular interest to oceanographers who are concerned with the shapes of the ocean geoid and the size and form of waves.

Other non-imaging radars are also referred to as scatterometers, a term usually ascribed to Moore of Kansas who states

'Any radar that measures the scattering or reflective properties of surfaces and/or volumes is called a scatterometer. Thus a scatterometer may be a radar specifically designed for backscattering measurements, or it may be a radar designed for other purposes such as imaging or altimetry, but calibrated accurately enough so that scattering measurements with it are possible' (Ulaby, Moore and Fung, 1982).

From this it will be seen that all imaging radars are scatterometers but not all scatterometers are imaging radars, thus here the concern is with the non-imaging types. The instruments usually house the transmitter and the receiver together and the returned signal is largely scattered rather than reflected back to the receiver, thus scatterometer is a preferred descriptor to reflectometer.

Scatterometers can be constructed to operate in different polarizations including circular polarization, they can also be made to operate over a wide range of frequencies to provide further target data.

Scatterometers may be airborne, used in space borne systems or ground based. In this latter mode they can even be hand held but are more usually mounted upon platforms such as cherry-pickers either as fixed platforms or, if truck mounted, as moving platforms.

Scatterometers are a major instrumentation in oceanographic satellites such as Seasat and the planned ERS 1 where they are used to provide valuable data on wind and waves.

It is in the research field that scatterometers are most widely used. Scatterometers allow precise measurements to be made of the backscattering properties of a wide variety of targets under varying conditions and at different frequencies and grazing angles. Studies over a period of time of a wheat field could for example assess the effect of row direction, of leaf size and angle, of grain head size and angle, of soil structure and moisture content under the crop and the effects of wind. These parameters can be observed for different polarizations, grazing angles and frequencies. Such studies are vital to gaining an understanding of the interaction between targets and systems to build up vital target data for eventual computer based image analysis systems and for optimum system design.

1.4 BASIC INSTRUMENTATION

The basic configuration of an imaging radar system has already been described in Fig. 1.2, each of these elements can have an influence on the eventual image rather in the way that a photograph can be influenced by lens characteristics, shutter speed, aperture and focal distance as elements of a camera system.

1.4.1 The antenna

The problems of the antenna have already been touched upon in the general description of radars. In particular the problems of the antenna size for a real aperture radar have been shown to be one of the main reasons why synthetic apertures are favoured. In studying the electromagnetic spectrum one is recording data as wavelengths and it is these differences in wavelengths that constitute the spectrum. It is also these differences in wavelengths that result in the various instrument design problems.

The microwaves used in radar are around five orders of magnitude greater than those recorded in the optical or visible segment of the spectrum. If the antenna is considered as the system lens, then optical lenses usually have diameters of $10^4\lambda$ scan related to the distance in $R/10^4\lambda$ in order to achieve satisfactory geometric resolution. It is generally accepted that for microwaves $60–400\lambda$ are considered realistic without recourse to mechanical scanning systems. If this is related to wavelengths of 10 cm the resultant antenna lengths become from 6 to 40 m. When the wavelength is related to the distance the problem increases so that in an airborne system, where the distance R is related to the flying height, the length of the antenna can be minimized and the aircraft selected that can accommodate the length of antenna necessary to achieve the desired image resolution. Aircraft performance is however related to size and to flying height so there remain a number of design constraints. In the case of satellites R is considerably increased so that the antenna required for a high resolution real aperture system becomes a practical impossibility. (One solution has however been suggested and this is described in Chapter 12.)

The design and form of the antenna has a fundamental influence on the transmitted and received wave pattern and form. To return to the analogy of the camera lens, modern camera lenses are capable of producing a virtually distortion-free image over the whole of the film area such that, for practical interpretation purposes, the image can be used without further processing. In most precision photogrammetric applications the lenses are regularly measured on an optical correlator to determine the distortion pattern which can be applied to any geometric measurements.

The radar antenna has similar characteristics and the design and construction of the antenna will have an influence on the returned signal and therefore on the image radiometry to give rise to the antenna pattern effect. It is possible, by detailed measurements on the images, to determine the antenna pattern for a particular radar; application of this compensating pattern to the image data can improve the overall image value.

The antenna is responsible for determining the width of the observed swath and for the across swath or width of the imaged area. Since the antenna is mounted rigidly along the platform, any tip or tilt movement of the platform changes the observed area of the antenna; an exaggerated roll of the aircraft could result in the radar imaging its far horizon and the sky for example, the oblique view intensifies any unstable motion effects. A stabilizer is usually fitted to compensate for minor in-flight motions and the use of large high flying aircraft further reduces the possibility of motion effects. One major advantage of the satellite platform over the airborne platform for radars is the increased potential stability.

This induced motion of the antenna may also result in the vertical pattern of the antenna across the image; as the antenna pattern will vary rapidly with changes in the angle of incidence the return signals will change, thereby causing variations in the image. Unless the interpreter has access to reliable flight log data it may be impossible to determine whether image irregularities are due to flying conditions or instrumental factors.

Because of its size, configuration and the requirements of reliable wave construction, the antenna is externally installed on the platform. In most cases an inert dome or housing is constructed to cover it, to improve its aerodynamic quality and to minimize air turbulence distortions and damage. Nevertheless damage can occur, particularly in RAR with their extra antenna length. Such damage or distortion can remain unnoticed until the effects are identified on the imagery, by which time it may be too late to correct operationally.

Finally the antenna is used both for the transmit and receive signals, thus a change in the antenna form can affect both signals which will increase the image deterioration effect.

1.4.2   The transmitter

The transmitter generates the energy necessary to provide a radar beam and transmits it to the antenna and thence to the target. The transmitted signal is an extremely short burst rapidly repeated to give the pulsed signal. In RAR systems a magnetron tube is usually employed which has a proven high efficiency at high peak power. For the SAR systems a different method is used, usually a travelling wave tube so that the phase can be fully determined. It is the phase difference between the transmitted signal and the received frequency which is used to determine the synthetic aperture.

The efficiency of the transmitter is closely affected by the power supply and any variations in the power supply can affect the transmitted signal and its pulsing. The most probable effect for the interpreter is a change in dynamic responses on the image, but system designs are constructed to minimize the possible effects and subsequent processing can reduce them further.

1.4.3   The receiver

The receiver is usually integrated with the transmitter, it receives the pulsed returned signal, determines the signal strength, relates it to the transmission times to calculate target distance and processes these data into a form suitable for recording. The return signal is at a much lower power than the transmitted signal since not all the signal is returned and only the refracted or backscattered signal is measured. It is therefore necessary to pass the signal through a low noise amplifier to increase or amplify the signal without adding further extraneous signals or noise. This amplified signal is itself further amplified through a mixer before measurement on a light intensity or a digital record. It is in the receiver that further signal correction is applied by a small computer unit, this compensates for the oblique view of the radar to present a near vertical view or the range correction.

Clearly any electronic instrumentation is affected by fluctuations in power supplies, normally these are prevented by the introduction of stabilizing units and a series of feedback controls. System effects from the receiver are not normally a factor which should concern the interpreter.

1.4.4 Recording

With RARs, recording is normally by direct image production. This is usually achieved by feeding the signal data to a modulating light intensity spot scanning across a cathode ray tube. Film passed over the tube at a rate related to the platform's forward speed and the ratio of the light intensity to film speed is used to record the image. The film is normally processed by a rapid on-line processor to present a continuous in-flight record of the images. Such an onboard processor can be seen in Fig. 4.1 which is the Motorola MARS system.

The problems associated with such a system are those normally associated with film processing. This is the only record of the image and therefore there is little possibility of compensating the images for any processing defects except through re-flights. Modern film manufacture minimizes the possibilities of variations in film performance either along one film or between films; pre-mixed chemicals similarly reduce inconsistencies in processing.

Automatic systems can suffer from inadequate maintenance and processing variations do occur. In an operational mode the film record is the only record and it is unlikely that an interpreter will be supplied with the original film data. Usually the original film will be copied either as prints or an intermediate film from which prints can be made. As with all film processes each successive process will result in some image degeneration normally in the form of a loss in dynamic range.

The film record produced from SAR imagery raw data is, by contrast, incomprehensible for interpretation and has to be further processed before it can be used, this will be considered in the section on processing.

Since RAR can be adapted to provide a digital record, this, like the film image, is the final product and is not normally subjected to further processing before release to the user. On the other hand digital data recording is the preferred method for SAR systems and is the only viable system for satellite platforms where data have to be relayed to the earth by a radio link. This digital record has to be processed to isolate the optimum signal return to provide a viable image, such processing is a highly specialized task.

One of the problems with digital radar recording of SAR is the extremely high data rate it produces. Recording is usually on high density digital tapes (HDDTs) and not the normal computer compatible tapes (CCTs). Since the data cannot be immediately used but have to be subjected to further processing, this will not affect the user who is usually supplied with processed CCTs.

REFERENCES

Moore, R. K. (1983) (Ed.) Manual of Remote Sensing, 2nd edn, American Society of Photogrammetry.
Ulaby, F. T., Moore, R. K. and Fung, A. K. (1982) Microwave Remote Sensing – Active and Passive, Vols I–III, Addison-Wesley.

2

System parameters

||

The system parameters are here considered to be those elements of radar as an imaging system which will affect the eventual image and are a function either of the instrumentation or of the system. There are further parameters which can affect the image caused by the target itself and these are considered separately in Chapter 3.

2.1 WAVELENGTH

The electromagnetic spectrum in Fig. 1.1 shows how much wider the microwave section is compared with the visible segment. At the same time it can be seen how each designated segment K, X, C, S or L may themselves vary in width and in some cases one individual segment can be broader than the visible segment.

In the visible segment the elements measured can be much finer; in broad terms the visible breaks down into the well-known rainbow bands but these in turn can be refined into shades of each band – the spectrum. In analysing an image each of these fine divisions can be a critical element in defining a so-called spectral signature of a target, it is the subtle combination of varying wavelengths of visible light that produces the conventional colour photograph.

This infinite multi-spectral approach cannot at present be applied to radar, it may become a consideration in the long term future as a result of considerable research but for the present imaging radars operate in a single band. It will be seen that even within these bands there is considerable latitude of frequency and that the signal and its response will encompass a wide 'spectral' range.

By reference to Fig. 2.1 and the table of bands and frequencies it can be seen that the well used X-band has a wavelength of 3.75–2.40 cm and a frequency range of 8000–12 500 MHz whereas the L-band has a wavelength of 30–15 cm and a frequency range of between 1000 and 2000 MHz.

Since radars dependent upon their own energy source and wavelength can be seen to be a factor of antenna design, it is not normal for a SLAR system to have a multi-frequency capability. It is of course possible to have more than one system mounted onto a platform each with its own antenna and for these radars to be focused to a similar swath path, but these are at present the exception rather than the rule. Most radars tend to be designed for a specific frequency or wavelength and it is essential to understand something of the respective attributes of the different wavelengths in order to be able to interpret an image.

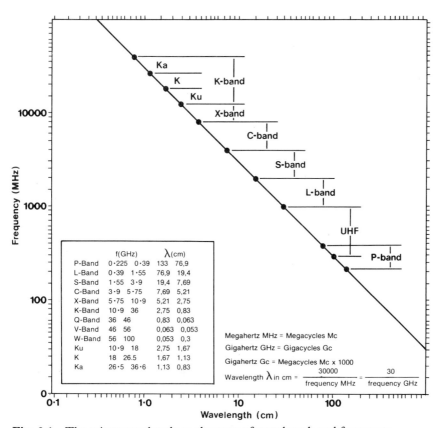

Fig. 2.1   The microwave bands as elements of wavelength and frequency.

The ultimate selection of the wavelength to be used may well be determined by factors unconnected with imagery. Radar wavebands are essentially radio wavebands in the higher frequencies, this means that certain wavebands cannot be used for imaging without the danger of interference with and from essential transmissive bands, usually of a military nature. At the same time radars are in use in numerous tracking systems for aircraft control, monitoring shipping and navigation and these bands must also be preserved. It is therefore necessary to be allocated frequencies by the International Telecommunications Union before a system can be designed and operated.

The selection of the radar wavebands to be used may be further constrained by the atmospheric windows. In Section 1.2 reference was made to the selection of segments of the electromagnetic spectrum in relation to these atmospheric windows, wavebands near the limits of the windows can therefore usually be avoided, especially for space systems.

Finally, the power available may preclude the use of very short wavebands, this is especially the case in current space craft where power availability may be restricted. Larger solar arrays and improved total system design may well improve the situation in the future but for the present these constraints apply.

The final selection will depend upon the type of application envisaged, Fig. 2.2 is a demonstration of the ability of different wavelengths to penetrate different soils. The selection of wavelength may therefore be related to the degree of surface penetration required.

The radar image is a representation of the amount of energy backscattered from a surface, the returned energy will vary with the degree of roughness of the surface. The wavelength used will affect the representation of this roughness with the shorter wavelengths generally giving a more pronounced roughness effect. In areas with an excessively rough texture the use of shorter wavelengths may result in a bright almost uninterpretable image, whereas in more smooth areas it may be essential to use the shorter wavelengths if any surface characteristics are to be expressed on the image.

As an example of the differences in requirements, a geologist working in an area of tropical rainforest may require the roughness resulting from a complex tree cover, to be suppressed in order to better observe the underlying land forms. The forester however may well find such imagery unacceptable on the very basis that the criterion used for discriminating between forest units has been removed.

The selection becomes more complex when the other major component is considered, the dielectric constant. The reflective ability of any target is related to its moisture content or the magnitude of the dielectric constant. A further effect is the amount of penetration that will result. The long wavelengths will have a greater degree of penetration, as will be seen from

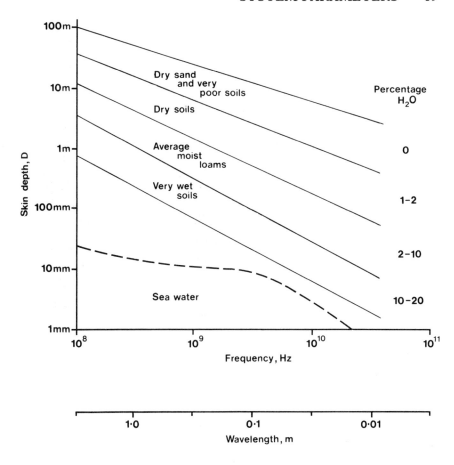

(Calculated from dielectric constant measurements)

Fig. 2.2 The relationship of frequency and wavelength to soil penetration.

Fig. 2.2, this in turn would result in the image representing the backscatter from a sub-surface component. This may be of value in cropped areas where it may be desirable to monitor soil moisture rather than crop state but may be less desirable in forest areas where an indication of crown size may be important. For a wavelength of for example 1 cm, penetration will be minimal, but the expression of roughness would be high for all surfaces; at the other end of the scale a 1 m wavelength could penetrate 1 cm in some dry soils but the expression of roughness could be negligible. A combination of wavelengths could offer advantages in estimating potential crop yield for example by enabling crop state to be assessed from one frequency and soil moisture content from the other after penetrating the crop.

Of the radars that have been operational one of the first, the West-inghouse, was a Ka-band RAR, this is no longer operational but many interpreters, particularly those concerned with vegetation studies, consi-dered this one of the best systems for interpretation.

Both the Motorola and the Goodyear systems are X-band and X-band has produced some of the most consistently useful imagery.

In the 1981–83 European SAR-580 experiment the system was designed so that X-, C- and L-band imagery was obtained, the X-band appeared to give a better overall data content than the other bands; however as the C-band was very much an experimental system the evidence was not conclusive. The waveband used for the only satellite radar in the civilian sector (Seasat) was L-band and the same configuration was subsequently employed in the shuttle experiments. These certainly produced some outstanding imagery with the SIR-A providing some dramatic observations on penetration in desert areas (see Chapter 11).

From the SAR-580 studies, experience with combining wavebands in a multi-spectral mode would infer that there are subtle differences between the images which in combination can produce important discrimination effects between units.

The foregoing only serves to emphasize an important factor in radar imaging, that there is a need for a great deal more experimentation and test imaging over a variety of terrains; such constraints should not detract from the very obvious existing benefits of radars and subsequent chapters will demonstrate how already viable benefits have been achieved.

2.2 POLARIZATION

The preceding section demonstrated how the selection of a different wavelength in a radar system can yield different data, this points towards the development of multi-wavelength systems. There is another development which can change the image aspect and help in discriminating surface features – polarization.

Most imaging radars use an antenna which generates a horizontal waveform, this is referred to as horizontal polarization. The observed surface breaks up the form however and some surfaces depolarize the image so that the backscattered signal may be both horizontally polarized and vertically polarized. The amount by which a signal is depolarized can be a further function of surface roughness.

By adding a vertically polarized receiving antenna to the system and linking this to another receiver channel, it is possible to identify and record these vertically returned waves. A switching mechanism in the recording system alternating between the two receiver channels would permit both the horizontal and the vertical returns to be recorded. Thus a relatively simple modification to a single waveband system will introduce a possible multi-spectral observational mode (Fig. 2.3).

As a further refinement the signal could be transmitted as a vertical polarization and received in the same mode or as a horizontal return. It is then possible to switch into all four modes to record alternately both the horizontal generated signals and the vertical generated signals.

When discussing polarization the notation HH is used to denote horizontal send horizontal receive with HV being horizontal send vertical receive, in a like manner VV and VH are also used. Note that HV and VH tend to produce similar results and normally only HH, VV and a cross polarization, whether HV or VH, are considered.

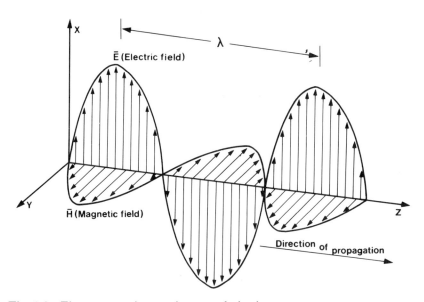

Fig. 2.3 Electromagnetic wave in two polarizations.

The observations on wavelength still apply and in the longer wavelengths much of the return may be subsurface scatter, there is some evidence to indicate that this in turn will affect the polarization in different situations.

There is little doubt that certain types of cover exhibit different image characteristics under different polarizations, these differences will vary in degree with different types of cover thereby providing a useful potential discriminant.

Radar waveforms need not be restricted to horizontal or vertical forms however and it is possible to generate a circular waveform and to receive circularly polarized images. Such techniques have been used in experimental applications and there are those who would argue that they can produce the best discriminating image. Such systems have not been used for SLAR type applications although they may become a future development.

The majority of systems in current use tend towards HH operation, however the SAR-580 referred to earlier did offer the possibility of VV and the related cross polarizations (Figs 2.4 and 2.5 give examples of HH and HV imagery).

There was some evidence that certain targets were better discriminated by use of HV or cross polarized imagery whereas between HH and VV there was little real evidence of preferred discrimination. The SAR-580 campaign however provided some evidence that the VV was giving an overall improved discrimination between distributive targets and an apparent improvement in resolution. In one case detail appeared on a VV image which could not be discerned on the HH image. Subsequent experiments have indicated that using two polarizations in combination can have a similar discriminating effect between units as the use of two wavebands, from which it can be deduced that polarization can be subtly different between otherwise similar surface targets and that the combination is necessary to identify these differences.

The ability to use some form of multi-spectral equivalent approach to improve interpretation is clearly desirable, the multi-wavelength method is one way of achieving this but it could be expensive to construct and operate. Multi-polarization working on one waveband may be a more cost effective alternative method of achieving a different look aspect.

2.3 RESOLUTION

The concept of radar resolution is usually one that is difficult for interpreters who are more used to examining photographs to accept. The definition best related to interpretation is that the resolution is the minimum distance apart that two radar reflectors need to be to produce two separate signals on the image. The term is not related to the actual size of the target, thus in a classic case with a bridge over a river, both the bridge and the river gave the same radar response so that the bridge itself could not be observed. However the street lamps on the bridge were at the optimum resolution distance and being excellent radar targets appeared on the image as strong points (Fig. 2.6). Thus objects can be below the resolution size and still be detected and they can be larger than the resolution and not be visible. Detectability is therefore not the same as resolution.

Many interpreters given imagery with what is regarded as a high resolution find it hard to accept that they cannot see important objects whose sizes are greater than the resolution factor, this emphasizes the need for

Fig. 2.4 SAR-580 imagery in HH mode (horizontal send horizontal receive).

Fig. 2.5 SAR-580 imagery HV mode (horizontal send vertical receive) of the area shown in Fig. 2.4.

(A)   **(B)**

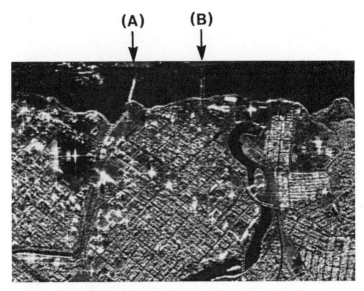

Fig. 2.6 The invisible road bridge. One bridge across the river (A) has a clear definition, it is a complex suspension bridge with many reflectors giving a total rough or bright target response. (B) however cannot be seen as a road surface, it is a smooth surface modern arch bridge. The bridge's position is only identified by the bright returns along each edge from street lamps.

interpreters to understand something of the system parameters and to re-think their logic of interpretation.

The above is the description with the most meaning for interpreters, however the resolution of a system is linked to the design elements of the system, that is the power, frequency, antenna design and look angle and to the speed of the platform. The definition of radar resolution in these terms is described as the half-power response width of the measuring system (Moore, Manual of Remote Sensing).

Radar operates on a pulsed signal which is beamed to a target and subsequently some proportion of that pulse is returned to the receiver where both the time lapse and the signal strength are measured. In one case the pulse width may be such as to hit two separate targets at the same time and the signals will be returned in parallel to give one strong signal return for both targets, the targets cannot therefore be resolved. In a second case pulse one may hit one target but the second pulse following close behind may hit the second target. The return from the first target is then still being received when the return from the second target begins to arrive, thus one continuous signal return is presented and again the targets are unresolved. Finally, pulse one may hit target one and the subsequent pulse may hit target two such that the return signal from target one has been received and a further non-response pulse returned before the response from target two is returned. The recording will now show two distinct target responses on the record, the distance between the targets to achieve this separate signal response is the resolution (Fig. 2.7).

In considering optical systems the resolution is linked either, in the case of films, to the grain size of the film used, or, in the case of scanners, to the dimensions of the electromagnetic sensors which give rise to the pixels of an MSS image. In both cases the sensor, whether it be the grain on the film or the EM sensor, can only record one averaged level of the intensity of light arriving at the sensor. Light from multiple objects therefore becomes averaged and the individual elements cannot be resolved.

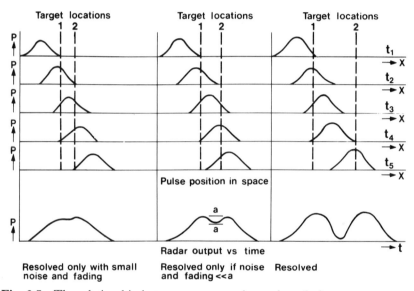

Fig. 2.7   The relationship between target spacing and resolution.

In radars there is a comparable resolution cell, this is an area over which the signal is averaged and recorded. Targets smaller than this cell cannot be discriminated although as with optical systems one small target of unique strength can dominate the response such that the signal return is virtually completely from that target. In optical systems bright light reflectors are often used as targets (usually mirrors); in radar, corner reflector targets have a similar effect and can appear as bright spots on the image, although the targets used may be smaller than the resolution cell.

In radar the cell is constructed from the angle of view which provides a footprint of the half power beam contour (Fig. 2.8), the range or the time

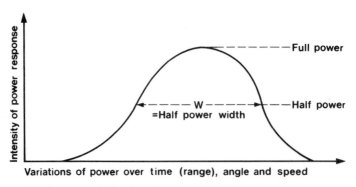

Fig. 2.8   Half-power width of a radar signal.

response and the speed of the imaging platform and these combine to produce a resolution cell.

If the platform is virtually static (as in the case of a cherry picker) and there is no differential recording over time, then the large footprint resulting from the look angle and giving a half power response is regarded as the resolution cell.

This will clearly only provide a target response of a wide area with very little discrimination, it may be useful however in trying to obtain a total target signature over a cropped area without resolving individual plants or being affected by crop rows other than in the overall context.

Where the platform is still static but a time lapse measurement is made to provide a range signal then the cell is narrowed to that area within the half beamwidth contour returned within a given time or at a given range (Ulaby, Moore and Fung, 1982).

In the use of aircraft or satellites for a platform there is however a forward movement of their platform to be considered. If there is no time or range measurement then the cell is restricted to a band equivalent to the variation in platform position between transmitting and receiving. In most operational systems, range response or time is included so that the cell now becomes that small area in the footprint of the half beamwidth that is within the same time recording span against the platform movement.

It is this final method of cell resolution using the half power response (Fig. 2.9) within the range and platform speed that becomes the Doppler beam sharpening or synthetic aperture radar. Clearly the smaller this cell can be made, the better will be the resolution of the radar system.

It is often said that RAR gives a better dynamic range, i.e. tonal levels, whereas SAR improves resolution and therefore one is better for vegetation surveys and the other for geology; in general however interpretation has to

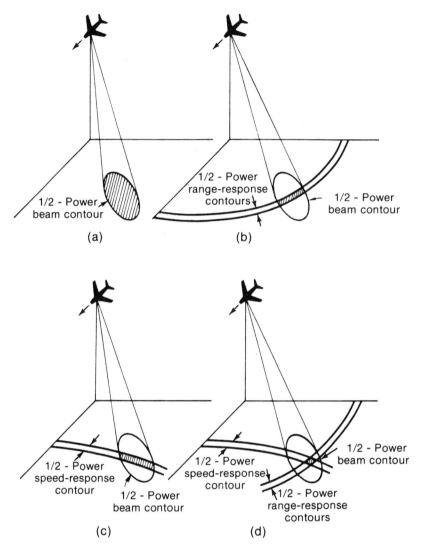

Fig. 2.9 Methods for resolution in microwave sensing: (a) angle only, (b) angle and range, (c) angle and speed, and (d) range and speed. (From Ulaby, Moore and Fung, 1982.)

be related to the total capability of the imagery and to the mapping scale.

At this stage it is probably pertinent to include some comments about scale as produced in radar. In cameras the scale of the imagery is a factor of the focal length of the lens against the aircraft flying height, with the general rule that the higher the flight the smaller the scale with the same camera. In radar the same rule does not apply, in the main commercial airborne systems, for example, imagery is normally obtained at 1:250 000 scale irrespective of flying height. The flying height can be selected to suit the terrain and conditions and the radar adjusted to produce imagery at the same constant scale.

With RAR systems fitted with real time film recording this means that photographic enlargement is used for larger scales with no improvement in the data content, only in the discrimination potential. In SAR where the data are recorded digitally the output scale is not a significant factor, it can be reproduced at any scale compatible with the total image data content.

In reality a number of other factors affect the possible enlargement of radar, there is for example the structure of the image itself manifested by the speckled representation of distributive targets, and the problem of resolution also restricts the enlargement factor.

Resolution and scale are nevertheless closely related. In optical terms the smaller the scale the more targets have to be aggregated into a cell and thereby the resolution is reduced. In radar much depends upon the ability of a system to record the vast amount of signal data being returned from an imaged area. There is, as a result, a relationship between the resolution cell and the data that can be recorded, which in turn will determine the effective image scales that can be used for interpretation.

There is a further determinant in resolution or the ability to discriminate targets and that concerns the recording of the tonal values. Two targets may be the correct distance apart to provide separate return signals but if the strength of those return signals and those of the background are the same then the targets will not be distinguishable.

This is one of the advantages of the electromagnetic optical sensors over photographic recording materials, the sensors have a greater sensitivity to variation in light intensity and can record many more grey levels than film emulsions. The number of grey levels is considerably more than the human eye can discriminate, therefore digital analysers are used to contrast stretch on the image and to enable subtle variations to be discerned and interpreted.

Grey level discrimination is equally important in a radar system, this can be a function of the instrumentation's ability to measure the strength of returned signals with sufficient variation to contrast a well segmented grey scale, and of the size of the resolution cell which is averaging responses to produce an average tone. Thus a system may be capable of very fine individual level discrimination but the resolution cell is so large that the averaging negates the value of this fine recording.

There is therefore a balance between the actual dimensions of the resolution cell and its grey level recording. As radar imagery is essentially a representation of surface texture there is a further factor involved in the grey level resolution, it will be affected by the texture element size of the terrain.

2.4 RADAR GEOMETRY

By definition any SLAR system is providing essentially an oblique view of the terrain, subsequent processing provides the interpreter with a pseudo vertical image but many of the factors connected with the oblique view remain. It is therefore essential to understand some of the basic factors in radar geometry and the construction of the image. Figure 2.10 demonstrates a radar imaging swath where the beam width is β. The depression angle v is the angle from the horizontal to the mid line of the scan, note this is not necessarily the mid-point across the swath on the image. The incident angle is the angle the far beam makes with the vertical. Nadir is the direct vertical or plumb point under the platform.

It can be seen that the angle of depression varies across the image and as a result not all targets are beamed at the same angle and in turn the return signal will also vary.

In Fig. 2.11 showing slant range and ground range, slant range R is

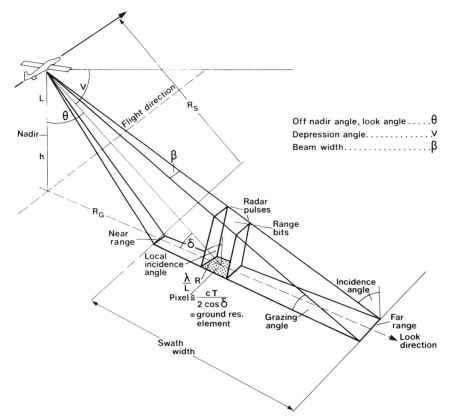

Fig. 2.10 The main elements of SLAR geometry.

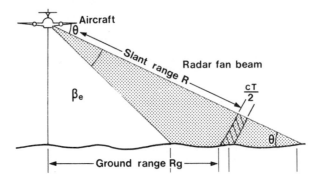

Fig. 2.11   Slant range and ground range.

defined as the radial distance from the nadir point of the aircraft or ground track to the target. The slant range is determined from the time *t* required for the pulse to leave the antenna and be returned to it, the slant range co-ordinates or position of a target are the system's natural measurements.

The system of using slant range has very few problems provided the terrain surface or features are flat, the problems occur in hilly country or when height related targets are involved.

SAR images are presented in one of two modes, either slant range or ground range. Slant range is shown in Fig. 2.11, the plane of the image is also shown in Fig. 2.12. Slant ranged imagery is related to a timed pulse to and from a target, on the other hand the ground range imagery is a factor of the height of the platform above the ground. In the case of an aircraft this is difficult to measure with precision unless something like a radar altimeter is used. When the ground is flat a constant factor can be introduced with reasonable reliability, when there is variable topography, without a precision profile recorder the height will be unreliable. In any case variations in the aircraft attitude will introduce further errors in the height

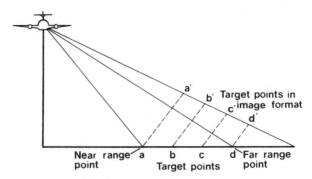

Fig. 2.12   Slant range image for flat terrain.

records since the profile will not be maintained along the nadir line. The use of a profile recorder would introduce a further complication in processing since the height factor would be constantly changing. It should also be borne in mind that the height profile only represents the terrain at the line of flight, the line of swath will be completely different in relief in such areas.

In the case of a satellite radar the ground height variation as a factor of satellite height is less pronounced since the satellite is at a much higher orbit. At the same time the height of the satellite will remain relatively more constant and stable.

2.4.1 Foreshortening

The occurrence of relief in the target area introduces a distortion known as foreshortening. The distance a–b, the top and bottom of a hill, projected onto the normal plane would be a–b″, however, projected onto the slant range image format it becomes a′–b′ which it will be seen is shorter than the true projected distance a–b″ (Fig. 2.13).

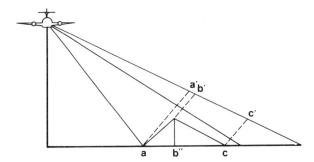

Fig. 2.13   Foreshortening.

From Fig. 2.13 it will be seen that all slopes facing the platform will be shortened relative to their true projected distance. The amount of foreshortening is related to the incidence angle of the radar, thus on an image slopes in the near range will be foreshortened more than those in the far range. Although an image may be corrected to give a pseudo-vertical view, the geometry of the image will vary across the image relative to the incidence angle and the relief of the terrain.

Table 2.1 gives the percentage of radar foreshortening against the incidence angle.

Table 2.2 demonstrates the relationship between terrain incidence angles against slope length for given depression angles where all the slopes in an area are 15°.

Table 2.1 Percentage foreshortening against incidence angle

Angle of incidence (degrees)	Foreshortening (%)
90	0.0
80	1.5
70	6.0
60	13.4
50	23.4
40	35.7
30	50.0
20	65.8
10	82.6
0	100.0

Table 2.2 The relationship between terrain incidence angles against slope length

Depression angles (degrees)	Radar slope length		Slope length ratio Back/Fore
	Fore slope	Back slope	
75	0	0.50	∞
65	0.17	0.64	3.76
55	0.34	0.77	2.26
45	0.50	0.87	1.74
35	0.64	0.94	1.47
25	0.77	0.98	1.28
15	0.87	1.00	1.15

A slope facing the radar will receive all the energy which would usually be spread over the ground projection area, i.e. a–b″. In Fig. 2.12, the return signals are now compressed into a shorter area a′–b′ with the result that the sum of the return signals produces a brighter image. On the back slopes the grazing angle diminishes so that the backscatter is reduced, therefore slopes facing away from the radar show a weaker return. The overall effect is that in hilly country slopes facing the radar will be brighter than normal and reverse slopes darker, the effect will increase with the severity of the terrain.

A further effect will be that the sum of the signals from the forward facing slopes will mask much of the texture resulting from surface features, making interpretation of vegetation cover difficult. The opposite effect on the reverse slope, a reduction in signal, will equally impair effective interpretation.

It will be seen from the foregoing that radar increases the visual representation of relief, this is one of the main attractions of the system to the geologist who is trying to map structures.

Figure 2.14 shows part of SAR imagery for an area of mountainous terrain at Luzon in the Philippines, the bright return of the forward facing slopes can clearly be seen, as can the darker back slopes. In this region the vegetation cover is uniform but no impression of vegetation can be obtained from the texture in the mountain areas. Some impression of foreshortening can also be gained from the image although only reference to a map would show the geometrical inaccuracies involved.

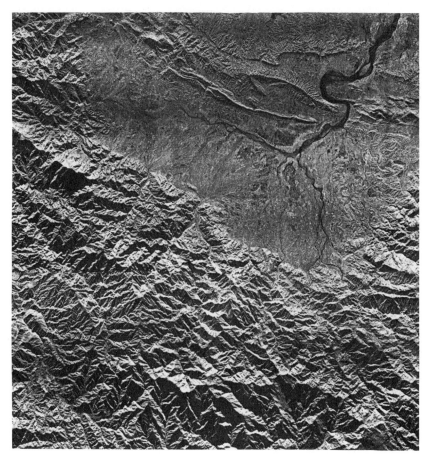

Fig. 2.14 Image of hilly terrain in the Philippines showing effects of slope. (Courtesy of Goodyear Aeroservices.)

2.4.2 Layover

When the radar has a steep angle and the terrain has extreme relief the foreshortening effect can increase to the point where the high point is imaged in advance of the base, such an effect gives rise to layover. This can be demonstrated by extending the figure used earlier for layover as in Figs 2.15 and 2.16.

The true projection of point b onto the base plane would be b″ with the distance shown as a–b″, on the radar plane, however b has given a return signal in advance of a so that the projected positions put b′ in advance of a′.

Radar layover is partially dependent upon the distance of the terrain from the radar instrument, on the slant range distance between the top and the bottom of the feature (which in turn is related to the slope of the terrain) and on the wavefront angle. Radar layover therefore tends to occur mainly in the near range of the swath and will decrease in effect towards the far range (Figure 2.17).

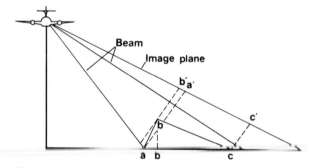

Fig. 2.15   Radar layover.

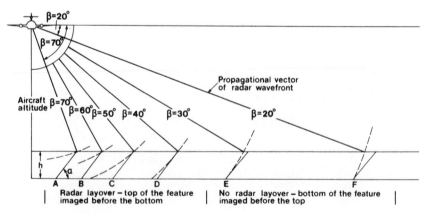

Fig. 2.16   Radar layover as a function of depression angle (β). (From the Manual of Remote Sensing, 1983.)

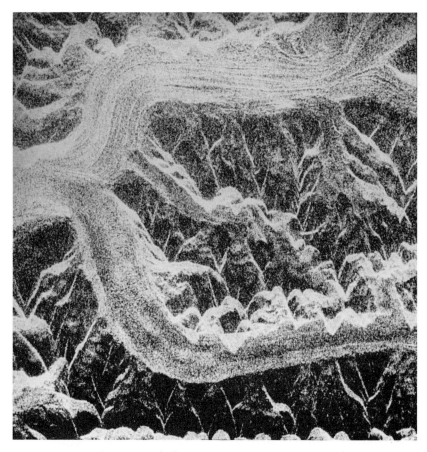

Fig. 2.17 An example of layover in an image, Seasat over Greenland. (Note where the peaks appear to overlap the glacier in the north.)

Correction, using slant range, within the instrumentation takes out the across swath scale effects that would result from an oblique view, but it does not remove this layover effect. Ideally it would be desirable to correct the signals further to relate to the true ground surface but this is not currently practical and in unmapped areas would be impossible.

An image that truly maps the observed terrain would have to be in ground range terms and not slant range. It has already been shown that ground range has to be related to the aircraft's true track position and for this it is essential that the aircraft is fitted with an accurate and reliable navigation system. This has to be linked to a reliable record of the aircraft flight attitude, since any change in attitude will affect the nadir and the look angle; with these data the effect of layover would not be eliminated, for this the absolute height of the targets would be necessary together with the height record of the aircraft relative to the true terrain height at the nadir point and these data are rarely available. Systems that overcome many of these criteria, particularly the effect of the compression, are referred to as true ground range (TGR) but these systems may introduce near distortion in high relief and it is always advisable to have access to slant range data as well. The effect is also clearly related to the depression angle of the radar system used and the position of the target in the across track swath, i.e. whether it is near range or far range. A comparison of a slant range display against a ground range display is shown in diagrammatic form in Fig. 2.18. The slant range display in mountainous areas may be visually preferable to the associated ground range display, in any case the interpreter would need to know what correction had been applied to an image and should have both versions available for comparison.

2.4.3 Radar shadow

Radar foreshortening produces a bright return on slopes facing the radar transmitter/receiver, this is one of the factors which produces the clear representation of relief on a radar image, the other is the radar shadow.

The radar shadow is an area of no data or total black, unlike an optical system where in shadow areas the light level is lower than in direct sunlight with the result that, unless the film used and exposure times are incorrectly adjusted, some detail in most shadow areas will appear on a visual image.

In Fig. 2.19 the principle of radar shadow is shown, this uses a

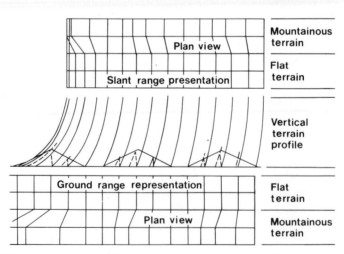

Fig. 2.18 Comparison of slant range and ground range display (from Reeves, 1975).

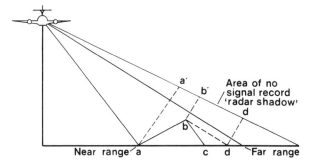

Fig. 2.19 The principles of radar shadows.

comparable demonstration to previous figures but in this case the beam hits the forward base of the hill slope a to produce a projected point a′, similarly all the slope is recorded up to the peak b projected as b′. No signal is received at the back slope base c, indeed no signal reaches the ground until the point d which produces the next return signal after b, between these points there is no record of a return signal and the time lapse record appears as a no response or black. This black record of the back slope to beyond the base produces the radar shadow.

The production of radar shadow is determined by the depression angle β and the back slope of the feature α_b. When the back slope is fully illuminated there is no radar shadow, that is when $\alpha_b < \beta$, if the two are equal ($\alpha_b = \beta$) then the radar beam will graze the slope, no record will appear of the detail on the slope but neither will there be a radar shadow. When $\alpha_b > \beta$, however, the back slope is lost and the return is from the ground in advance of the slope giving rise to the no data or shadow. As the depression angle decreases, as across the swath, so will the radar shadow of the same slope increase (see Fig. 2.20). The length of the radar shadow L can however be seen to be related to the slant range distance Ss of the target from the platform and the height of the target above the base plane h together with the slant range distance from the platform to the far end of the shadow Sr and inversely related to the radar platform altitude H to give a relationship of

$$Ss = \frac{L \, Sr}{H}$$

If the depression angle β is used this becomes

$$Ss = \frac{L}{\sin \beta}$$

It is therefore possible to make some statistical assessment of shadow areas to assist in classification.

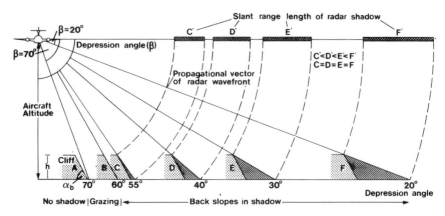

Fig. 2.20 Relationships of radar shadow geometry (slant range length) with depression angle (β). (From the Manual of Remote Sensing, 1982.)

In mountainous regions the large amount of shadow on the radar will result in large areas of unimaged terrain with an attendant loss of interpretive value of the image. It is for this reason that airborne surveys are normally conducted to give two sets of imagery from two opposing look directions, by this means area in shadow from one look will be in full view on the other flight path. The same effect will of course occur with radars imaging from satellite platforms but, unless two satellites were operating or the satellite was fitted with a dual direction radar, it would be difficult to achieve the same compensating action. Satellites usually operate in a near polar orbit which means they follow an inclined flight path around the earth. The satellite passes down one side of the earth and up the other side when orbiting, known as the descending and ascending modes, due to the inclined flight path; when as a result of the earth's rotation it passes the same target area on the different passes the flight paths will be at an angle to each other, the cross swath. In optical imaging satellites this effect is never seen since imagery can only be obtained while the satellite is traversing the side of the globe illuminated by the sun, and on the ascending mode it is in darkness. Since radars are not sun dependent, imagery can be obtained in both the ascending and descending modes and this change of look to some extent compensates for the loss of detail in shadow areas.

The radar shadow is therefore a stronger shadow than the one occurring on optical imagery and it is this which gives radar its clear expression of relief which is of particular advantage in geological interpretation.

In rugged terrain, circumstances can occur that will lead to a partial shadowing effect. This partial shadowing is imparted to the opposing slopes by the lower radar returns that result from radar beams at a lower angle of incidence. Slopes facing the antenna with higher angles of incidence will give a strong return signal and as a result a brighter image; as the angle of depression is lowered the liability to shadow increases. Clearly the effect of and amount of shadowing is critical in the interpretation of radar imaging for earth resources studies. Since the shadow is related to the depression angle of the radar the optimum angle to be used has evoked a great deal of debate. For the more subtle discrimination of target surfaces the higher depression angle is preferred, however the effect of a greater depression angle is to improve the target response in flatland areas but to be less acceptable in hilly terrain. Where imagery is to be used by a variety of disciplines some compromise is essential.

In nature the angle of rest of surface deposits is about 30°, this is the angle most unconsolidated materials will adopt, steeper slopes occur in consolidated materials but this is the maximum normally occurring in landscape. This being the case it is desirable for an imaging signal to be obtained from such slopes whether they face towards or away from the antenna. To meet this criterion a median angle of 45° is regarded as the most satisfactory depression angle.

Other constraints on the selection of the optimum depression angle are the power required by the radar, as power requirements will increase with imaging distance. In satellite systems the effects of the atmosphere and the earth's curvature have also to be taken into account.

It has been demonstrated that radar shadows are an important factor in interpretation, particularly in geological interpretation. The radar shadow is also related to the platform position, since the platform generates the energy, and not to an independent energy source. The effect of radar shadowing can be seen in Fig. 2.21.

If one considers a steep sided valley occurring in a straight section, then a radar imaging across the valley direction will beam along its length without any shadows being formed, the valley may not therefore be distinguishable as such on the image or at best may be assessed to be of lesser significance. In contrast a flight line parallel to the same valley direction will result in a strong shadow and a clear expression of the valley. The effect varies with the angle of linear targets to the flight path of the platform with the result that a linear analysis based upon strength of shadow can lead to completely misleading results if the directional bias has not been taken into consideration.

The look direction of the radar is therefore of critical importance in order to obtain the best possible linear representation of terrain. In satellite imagery the look direction is determined by the orbit paths and these are pre-determined for a complete global coverage, whereas aircraft flights can be planned to achieve maximum linear expression relative to the grain of the country. Where there is the possibility of influencing the look direction, as in

Fig. 2.21 Radar shadow accentuating relief on an image. San Cristobal Volcano, Nicaragua; imaged by the Westinghouse K-band system. A strong sense of topographic relief is obtained when viewed with radar highlights towards the top of the plate. If the plate is inverted, topographic relief appears to invert. (Hunting Geology and Geophysics Ltd.)

the case of airborne surveys, then the flight paths should be chosen taking into consideration the grain of the country in order to reduce the influence of radar directional bias. Figure 2.21 demonstrates how shadow enhances relief.

This linear bias is not however peculiar to radar, Landsat can be shown to demonstrate a similar bias. Landsat is sun synchronous and passes the same spot on the earth's surface at the same time on each overpass. There is a variation in the sun altitude however in accordance with the time of year. Comparison of multi-temporal imagery will help to minimize the possible effect on Landsat but this cannot be utilized with satellite radar. Figure 2.22 shows two histograms constructed from Seasat imagery in one case (a) and Landsat in the other (b); although the directions have changed a similar strength of linears in some directions compared to others can clearly be seen.

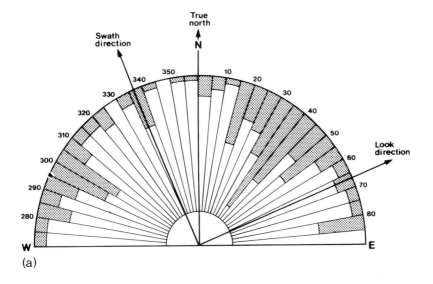

(a)

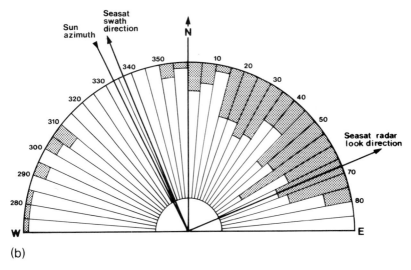

(b)

Fig. 2.22 A comparison of linear bias of Seasat and Landsat (from Hunting).
(a) Landsat lineament frequency and orientation.
(b) Seasat lineament frequency and orientation.

The diagrams were based upon the number and directions of linears identified by a team of geological interpreters.

REFERENCES

Moore, R. K. (1983) *Manual of Remote Sensing*, 2nd edn, American Society of Photogrammetry.
Ulaby, F. T., Moore, R. K. and Fung, A. K. (1982) *Microwave Remote Sensing Active and Passive*, Addison-Wesley Pub. Co. Inc., New York.

3

Target parameters

3.1 INTRODUCTION

In Chapter 2 we considered some of the aspects of SLAR which affect the image or influence the interpretability of that image. In this chapter we shall consider those aspects of the radar system which are more influenced by the interaction of the radar with the observed surface. In some instances it could be argued that the results belong to the system rather than to the target, to a certain extent it is difficult to make a distinction but nevertheless the character of the target has to be considered in relationship to the imagery source.

The radar energy sent to the target is reflected or scattered by that target. Some of the energy is absorbed, some reflected away, some diffused within the target and some eventually returns back to the antenna to be recorded and measured to produce the eventual image. The complete understanding of the scattering properties of matter is a subject for considerable research and study, in this chapter an indication is given of some of the problems and some of the current research being undertaken. In the end it is the interpreters who are concerned with the visual discrimination of images; the degree to which they can interpret an image depends upon whether they can identify a change in tone or texture as being due to system parameters (cross swath changes for example) or as being truly related to changes in vegetation or surface material. The extent to which interpretation can be made depends also upon the ancillary data the interpreter can use, i.e. his or her knowledge of the area and the parameters being investigated and knowledge of their ability to scatter microwave energy. In the latter instance a greater understanding of radar scattering may be desirable and the interpreter is referred to the numerous publications and studies available.

3.2 BACKSCATTERING

The ability to interpret features on an image is dependent upon the degree to which different targets will produce different signal returns in order to appear as unique features on the final imagery, coupled with the potential of the resolution of the system to discriminate between targets. Figure 3.1 presents the classical description of the manner in which surfaces affect the signal return to the antenna and are measured using radar. In the case of a smooth surface in (a) the energy is reflected away from the antenna and no return signal is recorded thereby providing a blank record or black image. With an increase in surface roughness, (b) and (c), the amount of reflected signal is reduced and there is an increase in the amount of signal returned to the antenna known as the backscattered component. The greater the amount of energy returned, the brighter the signal is shown on the image, so that white is the highest signal return. Radar imagery is therefore a measure of this backscattered component and since this is related to target roughness, radar is mainly a representation of surface roughness.

It follows from this therefore that targets will only be distinguishable as separate units if their surface roughness or backscattered component is different. Targets, which may in all other respects be completely different but with the same backscatter will be indistinguishable from each other on textural considerations and if adjacent will be inseparable. An example of this could be where a bridge crosses a river. If the water surface is smooth, as in the road surface, then the return signal will be equal and the bridge invisible on the imagery. In a classic example in Africa where villages consisted of grass roofed huts in a grassland area, the villages could not be discriminated.

As the microwave pulse will be beamed to a target at an angle related to the position of the target in the across swath, then there will be a variance in the amount of backscatter received, so that an otherwise homogeneous target will exhibit spectral differences across the swath with the variation in beam angle.

As the incidence angle increases so all surfaces appear to become less rough until at a 90° angle surfaces are apparently smooth, which returns, in effect, to the argument for side looking radars rather than vertical looking.

The representation of a target by its backscatter on the image is dependent upon a number of factors, the power transmitted from the antenna, the power actually received by the target, the ability of a target to backscatter the power, the loss of energy in the system, the introduction of energy from other targets and the amount of power received at the antenna. The power

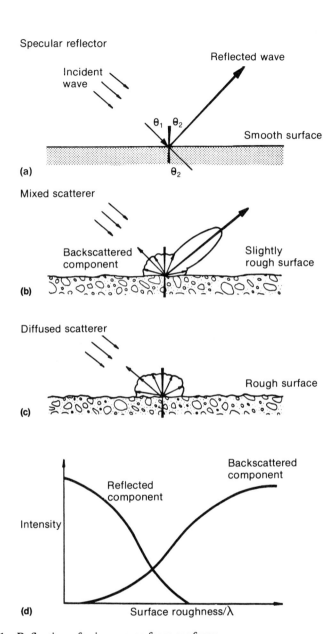

Fig. 3.1 Reflection of microwaves from surfaces.

received from other sources is referred to as noise and the relationship of noise to real signal power is known as the signal/noise ratio; each system is designed to keep this ratio to the minimum possible.

The power received P_r is calculated from the power transmitted P_t, the gain of the antenna G_t^2, the wavelength λ^2, the calculated loss of power in a system against the distance of the target $(4\pi)^3 R^4$ and the radar cross-section of the target itself $\delta°$ to give a formula

$$P_r = P_t \frac{G_t^2 \lambda^2}{(4\pi)^3 R^4} \delta°$$

The radar designer can control or calculate all of these factors except the radar cross-section of the target $\delta°$ and it is therefore this factor which is of vital importance in producing an image.

The apparent roughness of a surface will be dependent upon the incidence angle of the radar and its wavelength, thus a high frequency radar may record one surface as rough whereas the same surface viewed with a longer wavelength radar may appear smooth and the same surface across the swath may appear to vary in roughness with the grazing angle.

The flatness criterion was defined by Rayleigh. It states that a surface is seen as flat if the height difference between the top reflection as opposed to the bottom reflection is at the most $h = \lambda/8 \cos\theta$, see Fig. 3.2.

Cosgriff, Peake and Taylor (1960) demonstrate the variation in radar return with roughness for certain surfaces for varying incidence angles in horizontal polarization (Fig. 3.3).

The scattering coefficient of radar is referred to by the notation $\delta°$; in the

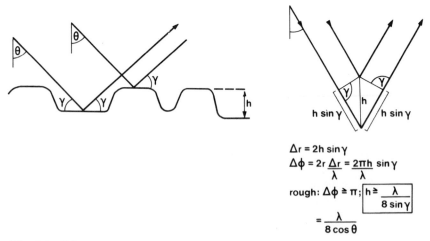

$\Delta r = 2h \sin\gamma$

$\Delta\phi = 2r \dfrac{\Delta r}{\lambda} = \dfrac{2\pi h}{\lambda} \sin\gamma$

rough: $\Delta\phi \geqslant \pi;\; \boxed{h \geqslant \dfrac{\lambda}{8 \sin\gamma}}$

$= \dfrac{\lambda}{8 \cos\theta}$

Fig. 3.2 Diagrams showing the definition of roughness according to the Rayleigh criterion.

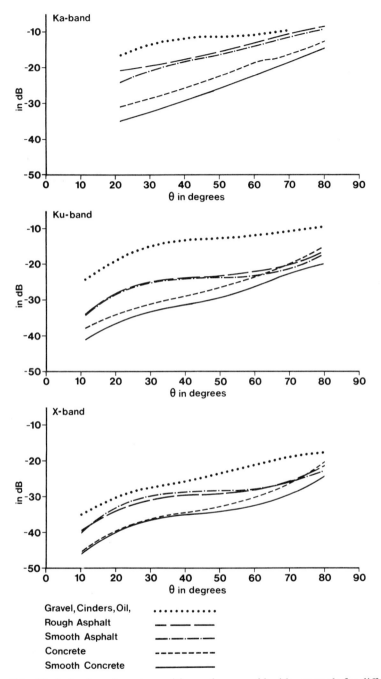

Fig. 3.3 Variation in radar return with roughness and incidence angle for different surfaces. (From Cosgriff, Peake and Taylor, 1960.)

design of a system the transmitted power, antenna gain wavelength and polarization are usually fixed so that the only variable is average return power strength which changes with the $\delta°$.

When a small incidence angle is involved the $\delta°$ value depends strongly on the slope. It is this slope sensitivity which enhances lineaments, drainage patterns, geomorphology and boundaries and is modulated into the image as a tonal variation. Where the incidence angle is between 15° and 70°, $\delta°$ is sensitive to small scale topography and to the material of the surface such as vegetation or surface moisture or, in oceanography, to small capillary waves.

When the larger incidence angle occurs there is a strong sensitivity of $\delta°$ to slope and strong shadowing from the geomorphology. At an incidence angle of around 15° there is a cross over in roughness scales so that various effects of surface roughness can be excluded, this angle is therefore of particular interest when determining soil moisture.

3.3 POINT TARGETS

The preceding description of backscattering refers to surface targets; however in radar there is another target, the point target. A point target is a discrete target with a simple configuration which gives a clear radar return, whose strength is disproportionate to its size. Figure 3.4 shows three common forms of man-made point targets. Spheres can also be excellent point targets and for this reason large raindrops can affect the radar return through clouds to give an overall bright signal return obliterating any other target responses.

Point targets of the corner reflector type are commonly used to identify known fixed points in an area imaged by radar in order to perform precise calibration measurements. Such targets can however occur naturally and are best seen in urban areas where buildings can act as trihedral or dihedral corner reflectors giving rise to intense bright spots on an image so typical of urban areas.

Usually most specular reflections travel away from the radar, since not many flat surfaces will be oriented perpendicular to the incident beam. However in an urban area where there are many right angles, successive specular reflections occur. For the particular case of three plane surfaces that are perpendicular to each other, the three successive reflections will combine to return the beam towards the antenna. Radar echoes from such areas tend to show extremes of contrast on the image. High returns (white on the image) originate from interior corners and surfaces perpendicular to the beam, but there is little or no return from most isolated surfaces such as streets, roofs, lawns etc. The configuration of buildings or the shape of the buildings themselves will often result in them forming excellent point targets to give high returns. A good point target can produce a dispropor-

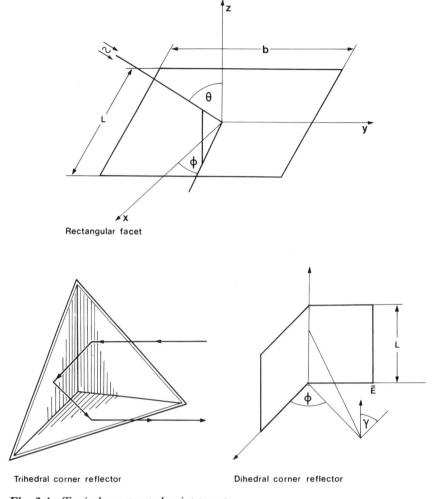

Fig. 3.4 Typical constructed point targets.

tionately large signal response as shown in Fig. 3.5 which also shows in graphical form the increase in signal return from a point target.

It is seen that the radar return signal can contain extraneous signals from other targets surrounding the prime target, therefore the ability to discriminate a target will depend upon its background. A poor target set amongst strong targets may be overpowered, while a strong target against a poor background may have the reverse effect to give a larger signal than its importance merits. This effect can lead to other confusions such as the identification of an edge between two otherwise distinctive surface targets.

Range ⟶

Azimuth ⟶

(a)

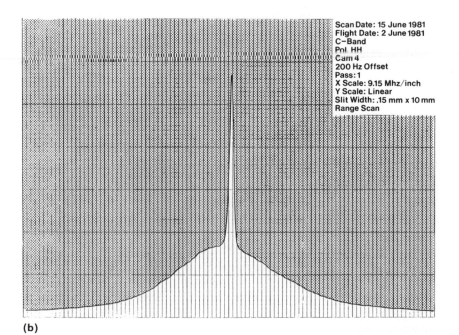

Scan Date: 15 June 1981
Flight Date: 2 June 1981
C–Band
Pol. HH
Cam 4
200 Hz Offset
Pass: 1
X Scale: 9.15 Mhz/inch
Y Scale: Linear
Slit Width: .15 mm x 10 mm
Range Scan

(b)

(c)

Fig. 3.5 (a) Point target response on an image and the resulting signal two dimensional spectrum, C-band system, 2nd June 1981, Pass 1, camera 4, polarization HH. (b) The azimuth signal response of the point target. Scan date: 15 June 1981, Flight date: 2 June 1981, C-Band; Pol. HH; Cam. 4, 200 Hz offset; Pass 1, X-scale: 9.15 MHz/in.; Y-scale: linear slit width: 0.15 × 10 mm; Range scan. (c) Radar image with distinct point targets, in this case these are trihedral targets set out for calibrating the radar.

An example of this often occurs in coastal areas where the land mass and the sea can be identified but not the boundary.

Point targets are an example of an object below the size of the resolution cell of the radar used that can dominate the return from that cell to make a clearly identified point. It can even give a strength of signal that will extend into surrounding cells.

3.4 VOLUME SCATTERERS

The simple explanation of backscattering given in Fig. 3.1 shows the effect from a surface, and most of the foregoing discussion assumes a return signal from the observed surface of a target. Unfortunately many targets, certainly those of vegetated surfaces, are more likely to be inhomogeneous and a vegetated canopy is a classic example of a volume scatterer.

The cover may be all trees, as in a forested area, which may be of different species with the attendant variation in leaf form and size, or grasses and bushes with variations in form, stalk size, leaf size and angle, fruiting and fruit size and finally an understorey of vegetation and a variable soil surface. Some of the energy will be backscattered from the vegetated surface, but some, depending upon the characteristics of the radar used and the target material, will penetrate the target and be backscattered from surfaces within the vegetation. Volume scattering is therefore dependent upon the inhomogeneous nature of the surface target and the physical properties of that target, i.e. leaf size, direction, density, height, presence of lower vegetation etc. together with the characteristics of the radar used such as wavelength and related effective penetration depth. When the radar wavelength is close to the dimension of the individual elements of the vegetative cover, for example the seed head of cereals or the dimensions of leaves, then resonance effects will occur and there will be higher $\delta°$ values. Figure 3.6 shows how tree cover and cereal cover act as volume scatterers. This scattering within a material need not be restricted to vegetation, in extreme arid areas such as the north African desert, where the surface material is dry loose sand, the radar can penetrate the top layer and be volume scattered within the particles.

3.5 PENETRATION

Volume scatterers rely on the ability of the microwave to penetrate certain cover targets and return signals from targets below that surface. In most cases the resultant signal will be a sum of the two effects in variable degrees

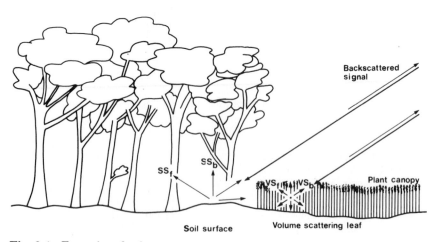

Fig. 3.6   Examples of volume scatterers.

depending upon the relative influence of the surface targets and the sub-surface targets.

In some cases the return from the sub-surface target can dominate the scene so that the recorded image is of sub-surface data and not surface. This may not greatly affect interpretation of, for example, vegetation, where it can be demonstrated that such a target signal can be related to an environment where a specified vegetation or vegetation mix is known to exist. In this case the signal can be identified as being the return from a certain vegetation unit when in reality it is from a surface below that vegetation.

In other cases the penetration ability can be of value in its own right; it could for example provide some indication of the volume of vege-tation, which in turn could be used for yield estimates; since the plant leaf growth is related to moisture content, study of penetration returns could indicate plant water stress conditions. At present these cannot be con-sidered as viable systems but they provide a field for considerable ongoing research.

On bare surfaces, penetration of the soil may occur and this may be an indicator of soil type differences, or, in a homogeneous soil, may be related to soil moisture content; both effects can occur simultaneously.

The electrical property of materials which determines penetration is the complex dielectric constant ε which is critical to radar return. The relative dielectric constant of a material consists of a real part ε' often referred to as relative permittivity and the imaginary part ε'' referred to as the loss factor to give $\varepsilon = \varepsilon' - j\varepsilon''$. The properties are very much dependent upon the water content of a surface so that penetration is greatest when there is a low moisture content and least when moisture content is high, the greatest reflectivity is in standing water.

There are obvious instances where this factor can be of especially interpretative benefit, soil moisture content has already been indicated, snow and ice studies are another example. The dielectric constant of snow is dependent upon the liquid water content within the snow. A great deal of study has been carried out on sea ice and snow where data on penetration phenomena are of economic importance. In sea ice the resulting image differentiation can be an indication of the relative age of the ice, whether it is recent or first year sea ice or older multi-year ice. Since snow is a mixture of ice and air the proportion of the two will indicate its density which in turn indicates its potential water content. Knowledge of this potential water content is clearly of value in studies related to run off, flood control and potential water supplies, thus the classification of snow areas in density units derived from penetration data is invaluable (Figs 3.7 and 3.8).

In geology the effect is less dramatic, in various studies carried out to measure the dielectric properties of various rocks it could be shown that the permittivities of rocks are related to their density.

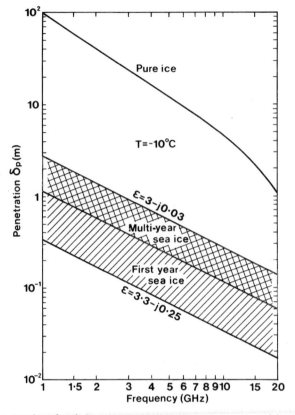

Fig. 3.7 Penetration depth in pure ice and in first year and multi-year ice (from Ulaby, Moore and Fung, 1982). Shaded areas correspond to the range of values of dielectric constant usually reported in the literature which arises in part from the range of salinities and densities commonly encountered in sea ice.

It is in vegetated matter that much of the research has been carried out in an attempt to derive values related to crops and crop growth state.

3.6 REFLECTION

The reflective characteristics of water (see Section 3.5) can also give rise to other strange effects on an image. For example a microwave signal reflected off the water surface and away from the antenna may strike another target above the water surface and from there be returned to the antenna to give the target a signal response greater than its normal target signature, see Fig. 3.9.

In some cases this can lead to a complex return of signals, particularly in the case of large high bridges as shown in Figs 3.10 and 3.11. Pulse A

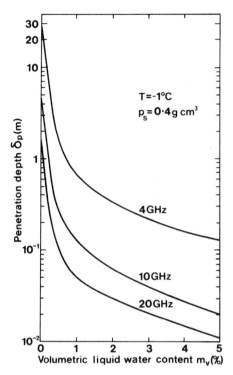

Fig. 3.8 Penetration depth in snow related to liquid water content (from Ulaby, Moore and Fung, 1982). The penetration depth decreases rapidly with the increase in liquid water content, especially in the immediate vicinity of $m_v = 0$.

reflects from the water to the top of the bridge superstructure and is backscattered from there to the antenna. Pulse B however reflects to the lower road surface of the bridge and is returned, while C reflects under the bridge surface and is lost. This phenomenon was observed in Canada on Seasat where the result was an image as a side view of the bridge itself. Such an action could also occur off smooth wet soil at the edge of a forest to result in a bright signal return at the forest edge. In a tropical swamp area this could result in a misinterpretation of the vegetation.

3.7 BRAGG RESONANCE

The sea surface is a special form of rough surface, it is for example in constant motion but may yield a uniform pattern related to climatic conditions, primarily wind strength and direction. Certain crop lands, notably cereals, can exhibit similar wind induced effects. There are two

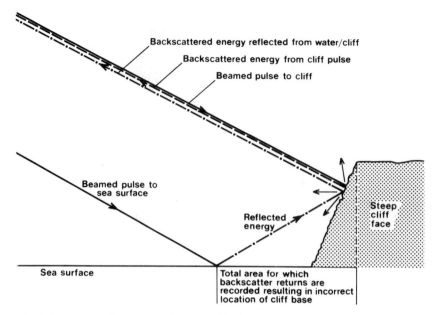

Fig. 3.9 Reflected energy backscattered by a secondary target. The beamed pulse to the smooth sea surface has been reflected away only to be further reflected by the steep cliff face, the beam is now backscattered to the receiving antenna to be recorded as a high return. The whole area from the sea surface receiving the first signal, to the top of the cliff face now appears white on the image.

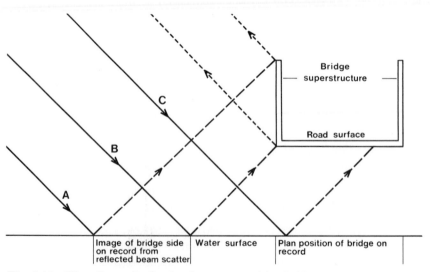

Fig. 3.10 The effects of reflection from water with a bridge.

Fig. 3.11   Radar image showing two bridges which appear with multiple images. Note that the bridges are in the along track direction and that other bridges at different angles do not demonstrate the same phenomenon.

types of waves for sea surfaces, the short capillary waves and longer gravity waves and it is the latter which are of particular interest to the oceanographer.

The influence of the wind on the sea surface inducing small capillary waves can be turned to advantage by analysing the backscatter from waves to determine the wind direction and strength. For this a special radar scatterometer is used, the wind scatterometer, a vital element in recent satellite and airborne radars.

The backscatter from a regular wave pattern can give rise to a resonance phenomenon known as the Bragg resonance (Fig. 3.12). Should the phase differences of the backscattered radar signals be multiples of the radar wavelength, then the reflected signals will summate and a strong signal be received. In Fig. 3.12 the incoming beam is coming in at the angle θ, the angle of incidence, the wavelength of the radar is γ and of the surface component is L. If the round trip phase difference between the returned signals from successive crests is 360° the signals add in phase, if ΔR is any other distance (except a multiple of $\gamma/2$) they are out of phase (Ulaby, Moore and Fung, 1982).

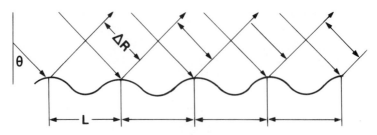

Fig. 3.12   Bragg resonance. In phase addition for those signals for which Bragg scattering conditions held: $\Delta R = n\lambda/2$.

3.8   CROSS SWATH VARIATIONS AND SURFACE ENVELOPE

Another way in which the radar signature can vary across the image is shown in Fig. 3.13 where an area of medium wooded savannah, otherwise uniform in aspect, in the near range emits backscatter from the tree canopy, the trunks and the ground surface; the returns from the ground and then the trunks will diminish across the swath until a return from the canopy alone is achieved in the far range.

To overcome these across swath variations, and indeed other across swath effects, flights are normally planned with a high degree of lateral overlap so that in interpretation only the centre of the swath is used. This is particularly important if it is intended to mosaic the imagery, the use of centre swath is the only means to achieve a reliable tonal match without using cosmetic photographic methods. Earlier the way in which the antenna pattern could affect the across swath dynamic range and how digital processing could be

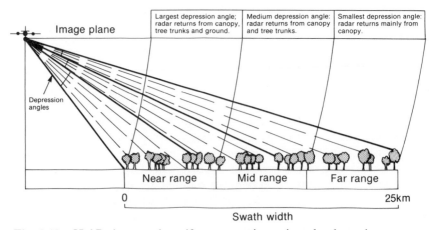

Fig. 3.13   SLAR signature in uniform vegetation unit on level terrain.

used to reduce this effect was described, it is worth noting that it is virtually impossible to make across swath corrections for signal variation of the same target due to the mixed signal returns described.

In contrast to slope and roughness which may be thought of as macro- and micro-scale roughness, respectively, there is a further important surface property which may be called meso-scale variation and is measured on the scale of the resolution cell. It can be thought of as the variability of the smoothed out envelope that would be formed by dropping a large soft sheet over the terrain, and in the case of natural vegetation it is equivalent to the shape of the canopy including any gaps there may be between individual specimens. It is roughness on this scale which accounts for the differences between adjacent resolution elements and hence the textural information in the image.

If there are many maxima and minima in the envelope in each resolution cell then the average scattering contribution from adjacent cells is very similar and the image texture will appear smooth or homogeneous. As the characteristic wavelength of the envelope becomes comparable with the ground resolution, so the return from each cell becomes more variable and the fine image texture more pronounced. For large scale envelope variations the texture will appear progressively coarser until it eventually becomes what would normally be called shape or pattern. Figure 3.14 indicates the mechanism whereby the meso-scale roughness influences the backscattered energy from each ground resolution element, though the precise mechanism is far from being completely understood.

In logged high forest, for example, the irregular canopy of large trees represents an envelope with characteristic features of the order of 20 m. Some range resolution cells will then have a large number of scatterers contributing to the received energy whereas others will contain only a few from the rear slope of the canopy or may even be in shadow. There will therefore be a very large variability between adjacent resolution cells, and the amplitude of the intensity variation on the image, perceived as texture contrast, will be proportional to the amplitude of the meso-scale envelope. It should be noted, however, that oversimplified statements to this effect in the literature can be very misleading. For instance, the envelope may have a very large amplitude for a plantation of mature trees but because the spacing is smaller than the resolution and the height is uniform there is very little textural variation in the signature. Since grassland or bare soil will also appear almost smooth on the image great care must be exercised when considering the relationship between the various parameters.

Much has been said about the radar shadow and the existence of a partial shadow. In Fig. 3.14 it is demonstrated how the image texture can be built up in a wooded area. It will be recognized that the resulting image will be a series of small alternate bright and dark responses caused by the canopy so

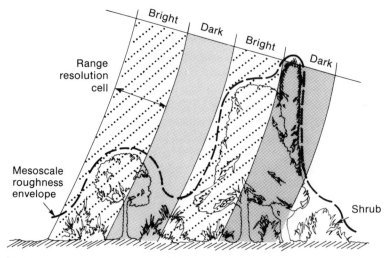

Fig. 3.14 Formation of image texture. If the amplitude of the canopy roughness on the scale of the resolution is large then there will be very different contributions to the scattering from adjacent resolution cells giving a corresponding variability in the image brightness.

that the build up of the image as the area is scanned will be of mottled or speckled appearance and not a smooth continuous grey tone.

A surface is said to be rough in the radar sense if its structure or shape has dimensions which are an appreciable fraction of the incident radar wavelength.

As with visible electromagnetic radiation, one of the major factors influencing the scattering properties of a target is its surface roughness. The main difference, however, is one of scale, so that whereas few surfaces are smooth compared with wavelengths of a few thousand Angstrom units, most natural surfaces have features comparable in size to the X band radar and most of the energy is reflected specularly (i.e. as light is by a mirror) and does not return to the antenna. Such surfaces appear dark or black on the image, whereas surfaces which are rough on the 3 cm scale will appear bright because they scatter energy almost isotropically (i.e. equally in all directions) and a certain proportion of this will return to the antenna. In the absence of slope variations, surface roughness on the wavelength scale is the other main factor, together with the dielectric constant, responsible for the average grey tone in the image.

The interpretation of radar imagery relies heavily upon the discrimination and identification of units by this roughness or speckle factor.

3.9 SPECKLE

At this point it becomes essential to make reference to speckle in more detail, it occurs as a factor of image production from roughness and from system factors. The radar engineers are continually seeking ways to reduce unnecessary speckle, some are aiming to produce a completely smooth image. The radar image is composed of a succession of resolution cells which can be likened to the Landsat pixel since the cell will comprise many scatterers at random ranges from the radar, the reflected or backscattered signals combine randomly giving rise to a complex interference phenomenon. The apparent reflectivity of a resolution cell will vary between the accumulation of all the amplitudes and zero.

The return signal strength in a radar imaging system is subject to fluctuations which lead to fading, it is these random fluctuations observed and averaged over a resolution cell or pixel which produce the speckle in an image. For an otherwise homogeneous target area some signals may return at full strength to give a bright pixel while fluctuations may lead to less return strength in adjacent cells to result in a darker pixel where the brightness is less than the δ° associated with the pixel.

The method of operation of a SAR can result in a number of observations or looks being made of a pixel; by examining each of the looks and averaging them it is possible to reduce the effect of speckle. With distributive targets such as a field of crops, there are a number of pixels covering the cropped area, by averaging the response from the pixels associated with a target it should be possible to arrive at an average value for the target.

Excessive speckle can result in an uninterpretable image and there are a number of ways in which speckle can be reduced in the processing of the image, unfortunately most of such systems tend to produce a loss of image resolution. At the same time a human interpreter can accept a great deal of speckle when carrying out optical interpretation, the brain will still recognize a whole pattern even though the scatter or speckle is random, and it will still recognize differences or similarities between units, thus this texture can be an important criterion in interpretation.

If analysis is to be by the use of digital analysis equipment, including interactive viewing systems, then this speckle presents a problem. The only method of differentiation is by tone, and standard density slicing will only yield a confused colour model which may show some coarse differences between units but cannot be used for quantified justification. A great deal of work has been carried out using low pass filters whereby the density level for a cell is averaged and the filter passed over the image to produce a balance equalized density reading. Whilst this does produce a smoothed image it unfortunately frequently destroys much that the interpreter was using as a visual discriminating factor. Surface roughness will produce a textured image, there is therefore a point at which the random or cell interference speckle is removed and the true surface texture remains. This point will vary with the overall surface roughness, thus a smooth surface should ideally exhibit no speckle at all while a varied height forest surface will exhibit a maximum degree of texture.

Research continues and some experiments using a mixture of low pass filters, linking of internal cells and edge detection are beginning to produce more encouraging results. If radar imagery is going to be gainfully merged with digital imagery from other sensors then such routines will be essential.

REFERENCES

Cosgriff, Peake, W. H. and Taylor (1960) *Terrain Handbook II*, Antenna Laboratory, Ohio State University, Columbus, Ohio.

Ulaby, F. T., Moore, R. K. and Fung, A. K. (1982) *Microwave Remote Sensing – Active and Passive*, Addison-Wesley Pub. Co. Inc., New York.

4

Image processing

4.1 INTRODUCTION

Radar image quality has been shown to be affected by parameters within the system itself and by the interactive properties of the observed target. The effect of these parameters can nevertheless be minimized or eliminated by the processing used to convert the returned signals into an acceptable image form.

The processing of SLAR data is a highly specialized task and in the case of digital SAR will require considerable computer power. Under normal circumstances the interpreter will have very little control over the processing and must needs accept the data presented as the best result that can be achieved from the processing.

In photo-interpretation interpreters will ideally have some knowledge of the photographic processes that have been involved in the production of the photograph (or image in Landsat) to be interpreted. Only through this knowledge are they able to recognize when a shortcoming in the image is attributable to the processing used or to recognize the extent to which it is possible to make spectral differentiations. In SLAR it is equally essential to appreciate the problems involved in processing and to recognize when further processing could improve the data.

Real aperture radars require less complex processing compared with synthetic aperture radars; in the RAR the target is accredited with one signature or backscatter return and the optimum signal does not have to be extracted from a collection of associated signals. Thus with RARs real time processing is usually carried out; in the case of a film record this will result in an immediate print out of the image, with a digital record the data as recorded require no further processing other than tape format conversion, usually from HDDTs to CCTs.

4.2 REAL APERTURE RADARS

The most obvious product from RAR is therefore an immediate image usually in the form of a print or film. Figure 4.1 shows the onboard processing used in the MARS radar system, the principle is shown diagrammatically in Fig. 4.2. The problems of any such output are that the system is constrained by the dynamic range of film which may be considerably below the potential dynamic range of the receiving system. There are some interpreters who argue that this is in fact beneficial since it introduces a filtering effect which results in a reduction of speckle or a smoothing effect on the image. Nevertheless if the intention is to carry out subsequent digital image analysis or to merge the data with other digital data then the image has to be digitized for acceptance into the system. The total related data loss may prove unacceptable for the intended studies. On the other hand a real time product of this nature is of considerable value where immediacy is vital, the study of sea ice conditions, monitoring natural phenomena such as floods or providing data on oil spillage for example. The

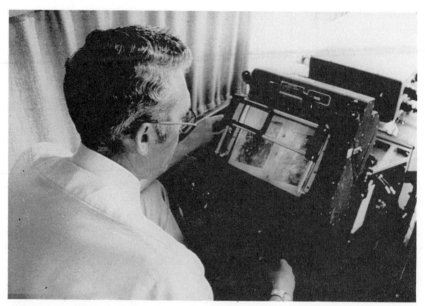

Fig. 4.1 In-flight processor for real time radar as used in the Motorola MARS system. (Courtesy of MARS Inc.)

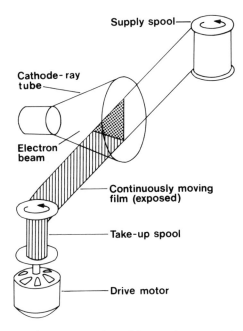

Supply spool

Cathode- ray
tube

Electron
beam

Continuously moving
film (exposed)

Take-up spool

Drive motor

Fig. 4.2 Diagrammatic representation of the continuous production of a radar film image. A light spot, projected onto the front of the cathode ray tube records onto film passing over the tube. The light spot scans across the film and is modulated to vary in intensity as the strength of the backscatter response increases or decreases. The speed of the film is related to the forward speed of the aircraft and the recording is continuous.

loss of dynamic range in these cases may not be significant when set against the need to receive and analyse data as speedily as possible.

From an operational point of view, access to real time processed data is of importance in ensuring the full survey area is covered satisfactorily and enabling gaps in an image to be identified and reflown.

Where only one product results and this is on a film base, then the preservation of this film becomes vital. For this reason the interpreter rarely has access to the original film but uses prints made from that film. Since each successive step in photography produces a further dynamic loss the data become progressively degraded with each photographic sequential step.

In some systems, usually of a lower order of imaging radars, the image is displayed in real time on a television monitor and this may be video recorded. The real time television display is ideally suited to situations such as oil slick monitoring; an observer in an aircraft watching such a display is only interested in identifying the presence of a slick, if there is no such slick then the sea image data do not have to be retained. Having identified a slick it may be necessary to direct a patrol vessel to the spot to arrange disposal, and for this the continuously changing display of the slick and the patrol vessel's progress when in range are ideal for controlling action. In some cases a direct transmission of the video image to the ship is made so that the captain on the bridge can also monitor progress.

Similar interactive real time systems can be envisaged for sea ice monitoring and monitoring physical disasters.

4.3   SAR OPTICAL PROCESSING

The method of recording SAR data onto film is not dissimilar to that used for RAR with the film passing over the CRT, however the result is not an image film as in RAR but a signal film or a film record of all the signals received by the SAR antenna. The film record contains interference fringes which record the phase information from a coherent radar detector to produce a type of hologram. In the case of the radar holographic interference fringes the interference is between the transmitted signal and the received backscattered signal. Figure 4.3 depicts the theory of hologram production.

The signal film without any further processing, is virtually uninterpretable to all except the most experienced eye and is never passed to an interpreter. For interpretation this signal film has to be further processed to produce an image film.

The final film products are produced in two forms: quick look or survey processed and precision processed. The former, as the name implies, is a means of achieving a more recognizable image which can be used to ascertain whether the survey area has been correctly imaged and whether the general standard of the image is acceptable, i.e. no obvious system induced faults have occurred. The interpreter is normally concerned with the results of precision processing, this processing has to be carried out on special equipment and will take some time to process after acquisition.

The precision optical processor is a fairly complex system built onto a rigid optical bench in most cases. Very few such correlators exist and this may present a problem in processing data, however when a commercial system is involved as in the Goodyear Aeroservices SAR then the contractor undertakes to provide image film as an end product. An earlier figure (Fig. 1.3) demonstrated how the image records of Goodyear SAR were built up; Fig. 4.4 is a schematic representation of their optical processor translating the film record into an image.

Another correlator is the one built at the Earth Resources Institute of Michigan (ERIM) known as the tilted plane optical processor, this is shown in schematic form in Fig. 4.5 and in the photograph of Fig. 4.6. In the ERIM processor the SAR data is introduced into the input place of the processor and illuminated with a coherent light beam. The light passes

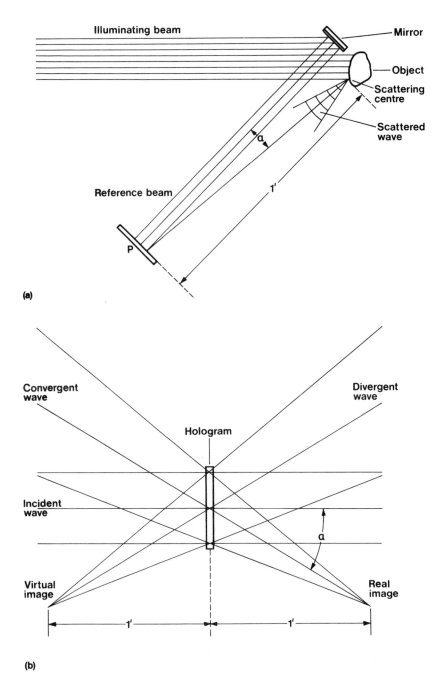

Fig. 4.3 The production of a holographic image (from Brown and Porcello, 1969). The top diagram (a) shows the conventional optical set up for producing a hologram. The hologram in (b) can reproduce both a virtual and a three-dimensional image.

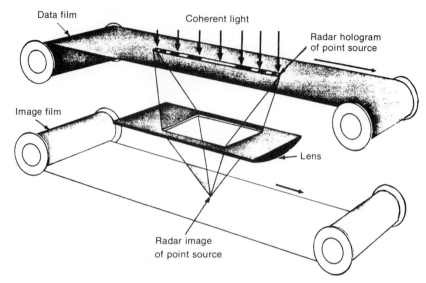

Fig. 4.4 Schematic optical correlator (courtesy of Aeroservice). Image reconstruction is carried out in an optical correlator which employs a cylindrical lens of the data record to produce point images from the complex pattern of reflected energy.

through the film image which has a full range interval of recorded data and a little more than a full azimuth length of point target history.

Since the signal phase term acts as a Fresnel zone plate, each point target signal history is self-focusing in both azimuth and range. The coherent light beam used to illuminate the signal film is focused when diffracted by the zone plate. Unfortunately the range and azimuth focal planes are not parallel and are separated. The azimuth focal plane becomes sloped compared to the range focal plane.

The telescopic lenses are a vital element in the filter plane processor and in this case two telescopic lenses are used in cascade, one being spherical and the other a cylindrical telescope. The spherical telescope re-images the light

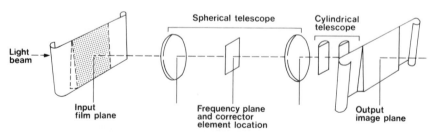

Fig. 4.5 Optical processor system schematic (courtesy of ERIM).

Fig. 4.6   The ERIM optical processor, raw data film and output film (courtesy of ERIM). The raw data film shows the holographic record referred to earlier, this is meaningless to the interpreter.

focused by the zone plates in the range dimension while the cylindrical telescope modifies the data in the azimuth dimension thereby removing the anamorphic property and adjusting the relative scale factors.

The illumination source is a laser beam and the system includes a number of correcting devices to remove or minimize other image effects. Only the optimum signal return reaches the recipient film to produce an intelligible image. The recipient film can be replaced by a video display to give a video image or by a photo-detector scan to enable the image to be digitally recorded onto computer tape.

It is obviously a much more complex and precise process than this description would indicate and like the majority of SAR processing it is relatively time consuming when compared to the processing of optical imaging data systems. As has been stated, there are only a few optical correlators in existence for radar images and these are mainly in the USA in research establishments.

The wavelengths of radar are considerably longer than optical wavelengths with the result that radar images usually have a highly specular appearance. One outcome of this is that the optical image before recording has a large dynamic range (> 60 dB), however photographic emulsions can normally only display 20–30 dB with the result that using films as a recording medium will inhibit the dynamic range of the radar that is recorded. The effect can be overcome by generating an optical hologram to be viewed on a holographic viewer, this will present the full dynamic range of the radar scene. To use this method the processor has to be converted from film recording to hologram production and it is not a system that is in operational use.

Precision processed optical SAR data are nonetheless of a high quality and are well suited to the requirements of most resources interpreters.

Optically processed imagery can only be produced at the imaging scale of the original film record, the correlator does not have any capability to enlarge the data. Similarly, although some geometrical corrections, such as range correction, are compensated for in the instrumentation, the errors described earlier will still be present after processing. In this event it is unlikely that the image could be used to prepare an overlay to an existing map or directly enlarged to scale. The use of a rectifying enlarger with a tilting baseboard may improve the geometry and provide a better general correlation with the map but the reliability with which this can be achieved would very much depend upon the amount of height distortion and therefore layover present. This condition occurs in normal aerial photography of course but the effects are increased in radar imagery. Anyone who has had experience of oblique aerial photography will appreciate the problem, especially if their experience includes the World War II Tri-metrigon photography, radar is still an oblique view and although range corrected, cannot be reliably rectified to topographic standards.

Map correlation and compilation may therefore present difficulties which would severely reduce the overall accuracy. Furthermore photography is an imprecise science with the result that variations in chemical mixture, temperature and timing can change the dynamic appearance of the processing rather than backscatter response.

4.4 THE DOPPLER PRINCIPLE AND PROCESSING

Earlier reference was made to SARs making use of the Doppler principle; before moving on to the electronic or digital processing of data it is necessary to consider this effect and its relationship to the SAR recorded data.

Standing on a railway platform and listening to an approaching train the pitch or note of the noise of the train will vary continuously as it approaches, reaching a peak at a point opposite the observer and then will continue to change as it passes away from the observer, this then is the simple classic description of the Doppler effect. In more scientific terms, if a source of electromagnetic radiation which has a fixed frequency of v approaches and/ or recedes from a sensor the sensor will receive radiation at a different frequency v' with v' greater than v when approaching and less than v when the source is receding. In SAR the target is passing through the signal beamwidth of the radar antenna and if the aircraft or platform is thought of as stationary with the target moving then the conditions for the Doppler effect pertain. In fact, of course, the sensor is moving relative to the target but the effect is the same.

In the Doppler effect the relationship between the frequency of the source radiation v and the received radiation v' is described as

$$V' = v \frac{(1-B^2)^{\frac{1}{2}}}{1-B \cos \Theta}$$

Where B is the ratio of the source velocity to the wave propagation velocity and Θ the angle between the direction of motion of the source and a line connecting the source and the sensor.

In the SAR case the radar emits a pulse at a frequency of v and the pulse received by the target is v' due to the forward motion of the platform, the backscattered energy is returned to the aircraft and is now the source while the antenna is the sensor but this latter is continuing to move relative to the source, thus the frequency observed at the platform is again shifted to v''. The change in frequency overall back at the sensor is now double the Doppler shift.

$$\Delta v = \frac{2v}{\lambda} \cos \Theta$$

This change in the signals phase resulting from the Doppler effect is the basis of the extraction of final data from synthetic aperture radars.

The Doppler system is used in a number of different radar systems, perhaps the best known to most people is the police radar used for the speed offence detection of vehicles. In other systems the Doppler is used for precision navigation and is a major contributor to precision flying for airborne surveys. Such radars are all continuous wave Dopplers.

For use with SAR systems a more complex type of Doppler is used, the pulse Doppler. It has already been shown how SARs use a pulsed signal (Z-), a pulse modulator being used to drive the pulse and the antenna accepting incoming signals between pulses; the incoming signal will vary with the Doppler shift. A sample of the carrier signal and a stable source are mixed and in order to remove unwanted differences (or sums) of frequency the signal is passed through a filter. The resultant filtered signal is mixed with the incoming Doppler shifted signal and passes on to the processor. This incoming signal contains all the data from a sweep view over a wide bandwidth, the data therefore have to be identified to give the range and the azimuth of a target as it passes through the sweep.

The processing of the radar return signal in the construction of a SAR image is based upon the assumption that the target is stationary, so that the relative motion between the target and the platform is known. In some cases however targets can have their own movement relative to the platform, such obvious examples being a railway train or a ship; both these objects are significant since they can be related to a known path, i.e. the railway line or the ship's wake, other targets can equally have their own motion as in an observed aircraft flying over the imaged area, but the true path may not be apparent and the final displacement may be less recognizable.

An object moving at v (velocity) towards the radar will have a Doppler shift of $2v/b$ relative to zero and its image will be shifted by Rv/V m forward (with respect to the radar velocity vector V) of its true position (Raney, 1982). In the examples quoted above the result on an image is that the railway line can be seen relevant to all other detail but the train will appear to be off the rails and running parallel to the track, a similar effect can be seen for a ship with the vessel appearing alongside its wake and separated from it (Fig. 4.7).

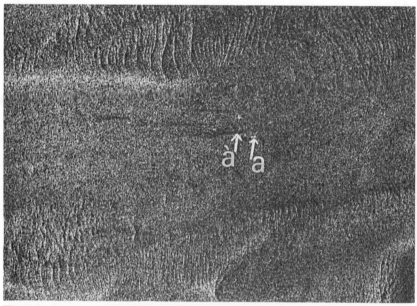

Fig. 4.7 Seasat imagery of part of the English Channel between Dover and Calais. The bright spot at a is a ship whilst a′ is its wake. Three other ships and their wakes can be seen in the vicinity. The amount of displacement can be related to ship speed and this factor is used to monitor shipping. Sea waveforms can also clearly be seen on the image and these in turn reflect the bottom topography of the area.

The topic of the Doppler principle is much more extensive and involved than is described here and it is not the intention of this book to go into too great detail on this and other related aspects. The average interpreter will not require greater detail but for those who find the subject worthy of further examination they are recommended to examine 'Microwave remote sensing' by Ulaby, Moore and Fung (1982).

4.5 ELECTRONIC PROCESSING

The on-board or ground based digital records, depending upon whether an airborne or a satellite system is used, record all the returning signals. One effect of this is the enormous amount of data generated by radar imaging systems and the rate at which those data are received. As a result data cannot be recorded onto normal computer compatible tape but have to be recorded onto high density digital tapes (HDDTs), using a high speed recorder. The large amount of data produced by SAR is one of the problems in processing the data and achieving a desirable data turn-round rate.

Digital processing seeks to produce a record of the optimum signal return assuming a look direction at right angles to the line of motion of the platform, to remove system signal artifacts and eventually produce a CCT which the user can reliably use for interpretation.

Despite the high data rate and the forward speed of the platform, the known pulsed waveform and its changes due to the Doppler effect etc. can be accurately recorded and subsequently identified, differentiated and notionally selected.

The radar signal pulse needs to be very short if it is going to achieve fine range resolution and yet has to be powerful enough to overcome any noise or extraneous signals. The pulse is therefore compressed in order that low amplitude pulses can be used rather than high amplitude pulses. The technique of pulse compression is known as chirp where frequency and pulse modulation are combined. The pulse frequency is modulated on transmission and the resolution obtainable is determined by the bandwidth over which the frequency has been swept and on the duration of the pulse.

In Fig. 4.8 the transmitted waveform shown in (a) is modulated in frequency from a low frequency to a high frequency, here the amplitude is constant throughout its duration. The 'de-chirped' waveform in (b) shows the $(\sin \chi)/\chi$ shape. The null to null width here is $2/B$ seconds but the effective width is approximately $1/B$. The amplitude is increased from 1 for the input signal to $\sqrt{B\tau}$ for the de-chirped waveform (Ulaby, Moore and Fung, 1982). The output of the waveform can lead to some ambiguity related to the actual position of the point target since the target may not appear as one distinctive dot but as a bright spot at the peak with fade off in each side direction producing side lobes. The production of the chirp and

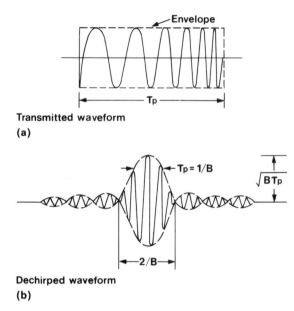

Fig. 4.8 Waveforms for chirp radar of a point target.

the processing is aimed at reducing the effect of this side lobing to a minimum.

It is the waveform of this chirp signal which is defined at the moment of transmission, and the effects of target return, time delays etc. on the chirp can be defined so that analysis of the return signals can result in identifying the optimum signal from the Doppler effect and associated range data. Unfortunately the transmitted signal form may not remain constant and in processing it is necessary sometimes to use either an idealized chirp, one that the system should transmit, or to extract from the data the actual chirp form used and to construct an averaged chirp for the survey data tape. Use of the wrong chirp in the calculations can produce undesirable effects upon the image by excessive side lobing or smearing where the image has a streaked appearance either across track, along track or both.

The radar is usually designed to have a bandwidth which will permit the Doppler spectrum to be divided up into equal parts forming a sub-aperture and for an image to be processed for each part. By adding the resulting images incoherently the speckle effect is reduced and there is an improvement in the radiometric resolution.

The digital bandpass filter used to generate a sub-aperture is known as a 'look filter' and these are designed to examine different segments of the Doppler spectrum. This filtering is normally carried out before azimuth correlation takes place and is referred to as pre-filtering. The pre-filter has to be so designed that it will not introduce illegal phase shifts into the data which would influence the azimuth correlation.

The number of looks used can be defined to give a multi-look filtering and the interpreter can define the number of looks required, provided the effect on an image is understood and recognized.

The processing of SAR digital data has therefore to examine the signals in order to extract the optimum Doppler return, to perform corrections for range and azimuth, to reduce the effect of sidelobing, to identify and suppress unnecessary noise and to eventually produce a new data tape that will contain the data necessary to produce the optimal image from a particular SAR system.

Reference has already been made to ground range and slant range correction, these are usually carried out in the processor and it is desirable that the interpreter is informed of the method used, the number of looks processed and other relevant data.

It is the necessity to read the data, process it onto disc, reprocess the data through another step, maybe store it again and reprocess, that makes the demands on the computer system both in time and space.

The final processed data can be output as a CCT or used to drive a precision photo-writer scanning with a modulated light source onto photographic film thereby producing a new image product for the interpreter. In this instance the film is not necessarily the definitive product as it is possible to return to the raw data and reprocess if unacceptable defects are noticed on the imagery. The CCTs are invariably used in image analysis systems where the image can be subject to further routines in an effort to improve their interpretative quality or they may be merged with other data or image sources such as maps or Landsat.

Table 4.1 lists some of the digital processors developed for processing Seasat SAR imagery and which have been used in some instances for SAR-580 data and for the shuttle imaging radar (SIR-A). These systems have mostly been improved and updated since the list was compiled but they serve to demonstrate the location of processing facilities and the complexity involved.

A major problem in digital SARs, particularly satellite SARs, will be the data rate that will be produced in imaging related to the short supply of processors and the time required to carry out processing.

Airborne systems such as the RAR of Motorola and the SAR of Aeroservice may not produce digital imagery or have the refinements of some of the more complex systems but for operational purposes they do have the current advantage of producing images within an acceptable time span. It is evident that for many aspects of resources mapping, particularly in the renewable resources, rapid receipt of imagery after acquisition is essential if the data are to be of operational value.

Table 4.1 SAR digital processors (after Guignard 1981)

	Configurations	Algorithm	Product	Throughput (hours)	Remarks
MDA Canada	32 bit mini + AP 120-B	Range Doppler (two-dimensional frequency domain)	40×50 km^2 25 m $\times$ 25 m^2 4 looks	7–8	100×100 km^2 available, sophisticated image analysis package exists
JPL (USA)	32 bit mini + 3 AP 120-Bs	Frequency domain	100×100 km^2 25 m $\times$ 25 m^2 4 looks	2–3	Operational facility
Communications Research Centre (Ottawa)	32 bit mini + AP 120-B	Two-dimensional filtering	21×47 km^2 25 m $\times$ 25 m^2 4 looks	$8\frac{1}{2}$	Variety of product, e.g. 7 m azimuth resolution 1 look
RAE (SDL) (England)	16 bit mini + AP 120-B	Convolution in time domain (azimuth)	50×50 km^2 25 m $\times$ 25 m^2 1 look	Very low	Experimental purposes only
DFVLR (MDA)	32 bit mini + AP 120-B	Range Doppler	40×50 km^2 25 m $\times$ 25 m^2 4 looks	7–8	Operational facility
CCRS (Canada)	DEC 10/90	Skew approach	100×100 km^2 100×100 km^2 16 looks	32	Experimental survey mode
Mitsubishi (Japan)	IBM/Cray	Frequency domain	20×10 km^2	$1\frac{1}{3}$	Software not completely vectorized
CCRS (Canada)	Interdata 3240 + AP 120-B	Range Doppler	40×50 km^2	8	MDA development (generalized SAR)
Norwegian Defence Research Establishment	General purpose	Frequency domain	45×55 km^2 25 m $\times$ 25 m^2 4 looks	50	Autofocusing
Nippon Electric Company (Japan)	General purpose computer	Frequency domain	40×50 km^2		Autofocusing

The problem is nevertheless recognized and since the launch of Seasat in 1978, when the problem of processing quantities of SAR imagery became apparent, rapid strides have been made in improving the performance of digital processors coupled with the parallel improvements in computer design and speed. There can be little doubt that the trend will continue towards digital processing especially as the use of digital image analysis increases.

REFERENCES

Raney, K. (1982) *The Canadian Radarsat program.* IGARSS, TP-6,1, Munich, June 1–4.

Ulaby, F. T., Moore, R. K. and Fung, A. K. (1982) *Microwave Remote Sensing – Active and Passive*, Addison-Wesley Inc., New York.

5

Radargrammetry

5.1 INTRODUCTION

In interpretation for resources the usual pattern adopted with imagery is to interpret directly from the image and to transfer the data to a base map; in certain cases, as with aerial photography, the images are placed together to form a mosaic over an area and interpretation is compiled onto or carried out on the mosaic. With multi-spectral data as from Landsat or airborne multispectral scanners, interpretation is aided by the ability to interrogate the individual spectral scenes and view them in a variety of combinations for subsequent comparative analysis. In these cases the digital data are used on an image analyser.

Under normal circumstances radar is single channel imagery although the possibility of a multi-frequency approach will be discussed later in this volume. Since it is single channel black and white the tendency is to treat it as a photograph and to process it in a similar manner.

In this chapter it is proposed to examine the differences between the two media for mapping and the errors that can occur.

5.2 MOSAICING

5.2.1 Photographic processing

The air photograph is an instantaneous image covering a specific rectangular area which will depend upon scale and camera format. The mosaicing from air photographs therefore requires laying down individual photographs which have to match four immediately adjacent photographs and subsequently either a map base or a series of fixed control points. In flat country the mosaicing can be extremely reliable but in hilly or mountainous terrain the radial height distortion will affect the matching of images and lead to duplication or omission of detail and will also introduce positional errors of photographs between control points.

Enlargement (or reduction) and rectification of each individual print can reduce the effect of these inaccuracies and with a skilled technician air photo mosaics can be made to a high degree of precision.

The radar image, whether RAR or SAR, is not a discrete rectangular image but a continuous strip record of one path of the platform used. Thus the larger strip should have the potential to eliminate some sources of errors (matching sequential photographs in a run) found with air photographs. The continuous record does however introduce its own error source, deviations in the aircraft flight or speed may make the air photography flight lines meander but the air photography only suffers a positional difference and not distortions as a result. With SLARs the film is a parallel sided record, deviation in the aircraft flight path will still record as a parallel strip, i.e. the film does not curve with the aircraft. Furthermore the film record is constructed from two principal components, a record of range either as distance from the platform or the nadir of the platform and an element related to the forward speed of the platform.

The quality of the imagery and its consistent geometric accuracy therefore depend, with an aircraft platform, upon the ability of the air crew to maintain a straight and level flight at all times, with satellites stable control is more assured. It can be seen that slight rolls of the platform will influence the location of the nadir point and affect range calculations, other variations in attitude such as pitch, the platform ascending or descending slightly, and yaw, where side winds cause the platform to be slewed at an angle while still maintaining a theoretical straight path, can all react with the image.

Aircraft used for SLAR are invariably fitted with advanced navigational and flight aids in order to help maintain a steady flight path, the use of high performance aircraft also helps minimize flight errors. Data on the aircraft's attitude are continually monitored and fed into the on-board pre-processor in order to adjust the recorded data for the observed attitude changes. Even so variations along images can and do occur with the result that the scale along track can vary with speed and the scale across swath can also be subject to variation.

With the air photograph an individual photograph can be rectified to reduce some of the distortion, with strip imagery this is not so easy although Motorola for example do operate a strip film printer which can enlarge or reduce a strip for mosaicing and make some attitude changes.

Radar images tend to be acquired at a fixed range of scales and not at specific scales for a particular application as with photography. This means that unless the imagery scale happens to match the desired mapping scale, some enlarging or reduction will be necessary in order to match the image to the mosaic control.

It has been shown how the side looking sweep of the radar will affect the imagery across the swath, so that comparable targets may exhibit variations in backscatter return in the near and far ranges. Similarly in hilly country the shadows will vary in the near and far ranges. The outcome is that matching parallel strips of imagery can be difficult in terms of tone and texture since one is matching a far range image to a near range image if one assumes flight lines have been flown as adjacent strips in the same flight direction.

Should the flight paths be in a turn and return mode (this applies only to aircraft platforms, where the aircraft images one strip, turns at the end of a run and images the second strip on its return) then the match is between two near ranges or two far ranges. Such flight patterns are usually avoided however because the images will result in reverse shadowing making matching even more problematical. The exception is where a dual look system is used, as for example the Motorola system, a turn and return pattern can be used since one is obtaining two look imagery in an effort to eliminate shadow. Figure 5.1 shows how such a flight pattern is built up.

In extreme mountainous country the problem of layover is introduced, thus the far range may exhibit little layover, but the overlapping strip, where the same area is now in the near range, may contain excessive layover making image matching for mosaicing almost impossible. In reality where the commercial airborne systems are concerned, for example, every effort is made to carry out internal corrections to the imagery to reduce the effects of

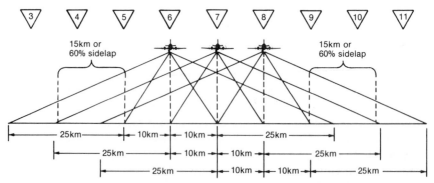

Fig. 5.1 The dual look imaging flight pattern (courtesy of MARS). It will be recognized that although the flight lines must be parallel, the flights can be made in alternate directs. Mismatch only occurs when the two antennae demonstrate different properties.

flight variations to an acceptable minimum. This has the result that in most cases skilled mosaicing by specialist technicians can achieve a surprisingly high degree of fidelity with the control data, whether those data be maps or control points. By obtaining adequate strip overlap and restricting mosaicing to the central area of each strip, the near and far range extreme effects can be kept to acceptable limits. In extreme terrain adjustment of the look angle and the flying height can help to compensate for the possibility of layover.

Dynamic imbalance between strips can still occur, this may be atributable to the radar system, usually as a result of antenna design, or may be variation in the photo processing. As with photographic mosaicing, the technician will attempt to achieve a cosmetically pleasing result, that is a mosaic where the identification of individual strips because of tonal differences is almost impossible. The methods used to achieve this are much the same as in air photo mosaicing, localized control of image development is used together with localized chemical image tonal reduction using reducing agents and finally some air brushing may be used.

The use of such techniques is acceptable and recognized practice and in the end a high quality even-toned mosaic will result that to the untrained eye has the appearance of a single image. For the interpreter the system may introduce unwanted interpretative defects, for example, tonal matching may have removed tonal/textural differences essential to an element of interpretation. Duplication and omission may still occur, although the technician will usually ensure this does not occur on recognizable features, this will however influence any area measurements that may be made. For these reasons the interpreter is strongly advised to carry out interpretation on reproductions of the original imagery and to use the mosaic only as a regional compilation base.

One distinct advantage of the radar mosaic will occur in geological interpretation where analysis is made of linears. Accepting that radar has a linear directional bias, radar will already improve the representation of relief, and the use of strips as opposed to small rectangles in a mosaic will minimize the possible occurrence of local mis-orientation of images which could seriously influence the directional pattern of linears. Figure 5.2 is an excellent demonstration of the sequence involved in the construction of an airborne radar mosaic.

Satellite radar obtains imagery in successive parallel passes and is mosaiced in much the same way, the relative stability of the platform will minimize the perturbations due to platform attitude and the height of the platform will minimize the relief distortions except for the most severe terrain. Furthermore the speed of the platform will be relatively constant. Experience with Seasat has shown that a reliable mosaic can be constructed from the data and mosaics were successfully made of the major part of the

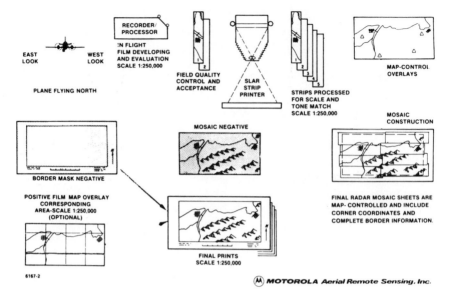

Fig. 5.2   The photographic mosaic schematic.

United Kingdom and of Iceland which compared favourably with the base maps at the same scale.

5.2.2   Digital mosaicing

The foregoing has been concerned with mosaicing from the photographic product. It has already been discussed how the preference is for digital recording. The digital record can be turned into a photographic image form and during the digital processing stage the grey level histograms for images can be matched to produce images which will be comparable in tone. Mosaicing will therefore be comparable to that carried out from the photographic record base with the added advantage that scenes can be matched without artificial loss of data content. Finally images can be mosaiced by completely digital methods, that is scenes can be warped to fit identified control points by the computer, strips can be fitted to each other, the dynamic imbalance adjusted to a common level until finally a new data record is made of the reconstructed or mosaiced image which can be reproduced in total using a photo-writer. The constraint may be in the film format that can be accepted on the plotter, many plotters or photo-writers will only produce an image of approximately 25 cm square. Larger plotters are available but the smaller size is more usual. This means that the image has to be enlarged to the final scale with a consequent loss of detail.

Photographic mosaics by comparison are laid at imaging scale and photographed on large format copy cameras to produce a scene at compatible or smaller scales, and when reduction is involved this results in an increase in detail relative to the final scale and not a loss of detail as in enlargement.

In a recent study involving the quantification of Seasat imagery mosaics four adjacent passes of Seasat over Alaska were digitally mosaiced. The scenes were radiometrically balanced using histograms constructed from data in the near edge and far edge of the image swath using 2 km × 100 km segments. A smoothed linear stretch was performed based upon the difference in mean and standard deviation between the histograms. After rectification the frames were reduced to produce imagery approximately 2000 × 2000 pixels spaced 50 m in both azimuth on range. The resolution was reduced to 100 m with 64 looks to reduce speckle and the final mosaic was found to have a registration error of less than one resolution element (100 m). The area covered by the mosaic was approximately 300 km × 150 km. In this study no tie points were identified between adjacent strips and the position of the strips relative to the orbit was accepted in order to demonstrate the potential of constructing digital mosaics from spaceborne imagery as a method of providing global mapping.

5.2.3 Multi-temporal imagery

Airborne surveys are usually conducted to a contract specification whereby all the area is covered within a relatively short period of time. In general this will mean that differences due to changes in target aspect between flights will be minimized and therefore image matching will not be severely affected.

This does not always hold true for radar since it is well known that radar backscatter is related to the dielectric constant, any changes in this constant between flights can produce images with different backscatter representation. A flight plan may start during a dry spell for example and conclude with the occurrence of the wet season, it can be realized that this change in soil moisture content will in turn change the appearance of the image. When a flight plan, after checking, shows gaps in the desired imagery and these are filled by re-flights in different climatic conditions the new imagery may show distinct differences when compared with the original imagery and be difficult to reconcile in a mosaic. Such an instance occurred during the survey in Nigeria, where in the northern area, dust storms curtailed the flying programme and the aircraft was relocated to another area in the country, returning later to complete the imagery, at which time the seasonal rains had arrived. The result was a strip of imagery with such marked differences that for a while the interpreters feared they had negatives rather than positives.

When considering satellite imagery the possibility of using multi-temporal imagery increases. Although subsequent adjacent passes may be made on successive days there are constraints imposed upon obtaining imagery derived from the shear volume of data involved. This could well mean that obtaining imagery for an adjacent pass may have to wait for a further orbit, during which time weather changes could well alter the backscatter response of soil/vegetation targets. The digital adjustment of mosaics in an automated mode means that radiometric corrections based upon comparative histograms could, as a result, produce a completely unacceptable match.

Other multi-temporal changes can also occur, for example crops may have been harvested in the time delay period, natural disasters such as flooding could occur or seasonal rain could change river courses in arid areas.

Any attempt to operate an automated mosaicing system that does not take cognizance of such potential differences could well lead to erroneous mosaics and biased data being presented to the interpreter. The safe way therefore is for the interpreter to have equal access to the strip imagery.

5.3 STEREOSCOPY

In the interpretation of conventional aerial photography, the use of opposing target views obtained from pairs of photographs to produce a stereoscopic effect is an essential aid. The production of precision topographic maps with reliable height data and contours from aerial photographs has been developed from the stereoscopic viewing and measuring of such images.

It is perfectly feasible to fly radar in such a way as to obtain overlapping images to give the possibility of stereoscopic study. However radar stereoscopy is not the same as photographic stereoscopy. Radar is an oblique view corrected to give the geometric appearance of a vertical view but the effects of the range and shadow have already been discussed and it is from these effects that the problems occur, particularly from layover.

In discussing the effects of radar shadow the use of opposing views was shown to be a possible solution to the problem, these opposing views can be used to give the desired change in look position and in stereoscopic observations. Unfortunately the reason for the opposing look direction, namely shadow effect, gives rise to one of the difficulties in stereoscopic observation.

Since in a steep object detail in shadow is lost, the opposing view is used to image the lost detail, but this results in a shadow over the previously observed data. Detail on shadow slopes is therefore only observed once and only the top of the target observed from two directions, i.e. the hill or mountain ridge is capable of stereoscopic observation. Detail in valley bottoms may well occur in shadow from both looks thereby eliminating the other valuable height reference base.

Layover, whereby signals from the top of a mountain may be received before signals from the bottom or along the facing slope, presents another problem. The imaged area beyond the representation of the mountain top will therefore be made up from signal returns from lower down the slope which are received later than the returns from the top and thus are deemed to be further away (Section 2.4.2). The resulting layover effect leads to confusion for the observer when the image is viewed stereoscopically. In certain circumstances, as for example an image of an isolated pointed peak, there will be a layover effect from the opposing views required to produce a stereoscopic image and, either under a stereoscope or in a photogrammetric measuring instrument, it is impossible for the interpreter to obtain a stereoscopic view from the merging of the two sets of layed over data.

The effect of opposite look direction is shown in Fig. 5.3 (Ulaby, Moore and Fung, 1982) where the height of the object H is observed from two positions 1 and 2. It has a layover of distance ΔR, and ΔR_2. The effect when viewed under a stereoscope is to give a three dimensional view but the displacements are opposite to those observed in air photographs.

Parallel flight lines in the same look direction can be planned to overlap and this results in another potential change of eye position required to achieve stereoscopy. In this instance some of the effects of shadow are overcome in that one slope is viewed twice and only shadow details remain obscured. Layover however remains a problem. Figure 5.4 (Ulaby, Moore and Fung, 1982) demonstrates the geometry of this same side viewing where in this instance the difference of ΔR, and ΔR_2 determines the height of the viewed object.

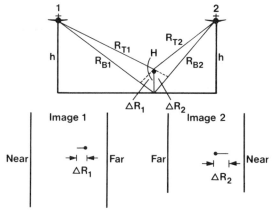

Fig. 5.3   Opposite side stereoscopy (after Ulaby, Moore and Fung, 1982).

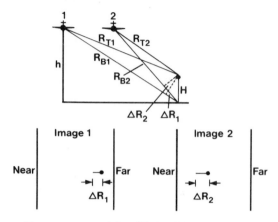

Fig. 5.4   Same side stereoscopy (after Ulaby, Moore and Fung, 1982).

Such flight paths are easier to establish in satellite imagery than opposing looks and give a preferred stereoscopic image, there is therefore a possibility of using stereoscopic measuring from satellite radar imagery.

The problems of the reconstituted oblique view and stereoscopy will have to be overcome before radar can be considered as a precision mapping tool. In reality the inability to use radar for stereoscopic examination has not proved to be a major drawback to most resources studies. The improved expression of relief in radar through intensified shadowing to some extent reduces the problem and enables the interpreter to map the geomorphology and at the same time make some relative assessment of height even from targets such as edges of woodland and crops.

The variations in the platform attitude or direction of movement between flight lines, whether same side or opposing looks, which affect the overall geometry of the imagery in the oblique view, and which were discussed in mosaicing, would also add undesirable distortions to the stereoscopic image.

Early work carried out by Leberl in 1976 (Leberl, 1976) indicated that in areas of moderate to low relief a planimetric accuracy of 100 m and a height accuracy of 20 m were possible. Leberl also compiled some tables (Tables 5.1, 5.2 and 5.3) (Leberl, 1983) which demonstrated the results obtained on stereoscopic examination of images by various researchers and comparable results on planimetric accuracies from different researchers using various SLARs. Although these studies are being pursued and have been taken further the tables still provide a useful guide to the radargrammetry potential of SLARS.

Subsequent studies by Schanda (1984) using SAR-580 imagery in a mountainous region mapping at scales of approximately 1 : 140 000 using a stereo comparator were analysed to a sensitivity of 1 μm. The conclusions

Table 5.1 Evaluation of viewability of a large number of stereo radar images (after Leberl, 1983)

Type of radar	Number of models studied	Base length (km)	Look angle Ω (degrees)	Type of stereo	Intersection angle ΔΩ (degrees)	Type of terrain	Stereo viewability
Seasat-SAR	10	25–75	20	Same side	1.2–4.8	Rugged	Very convenient
		550	20	Opposite side	40	Rugged	Not possible
Aircraft SAR	4	0.7–13	68	Same side	0.2–23	Rugged	Very convenient
Goodyear	2	30	68	Opposite side	120	Flat to rugged	Only when flat
Aircraft real	1	10	81	Same side	6	Flat to hilly	Convenient
Motorola	1	48	80	Opposite side	160	Flat to hilly	Convenient
Lunar Apollo 17						Flat	Convenient
ALSE-SAR	19	0.7–10.3	10	Same side	0.3–53	Rugged	Only with ΔΩ <119

Table 5.2 Radar stereo accuracies obtained by various authors (after Leberl, 1983)

| Source | Year | Accuracy (m) | | | Control per 100 km² | Antenna | | | | Remarks |
		Along 1σ	Across 1σ	Height 1σ		Resolution (ground-meters)	Stabilized	Type	Radar system code	
Gracie	1970	12.2	7.7	13.2	35.0	17	Yes	Synth.	AN-APQ 102	Opposite side
		68.0	138.0	240.0						Opposite side
Konecny	1972	130.0	428.0	1548.0		17	Yes	Real	Westinghouse	Same side
		26.8	21.9	16.7						Opposite side
DBA-Systems	1974	29.5	25.6	19.7	1.2	3	Yes	Synth.	AN-ASQ 142	Same side
Goodyear	1974			93.0	3.3	12	Yes	Synth.	GEMS 1000	Same side
				33.0						Opposite side
Derenyi	1975			177.0		12	Yes	Synth.	GEMS 1000	Same side
Leberl	1975	173.0	510.0	109.0	0.3	20–150	No	Synth.	Apollo 17	Same side small base satellite

Table 5.3 Single image radargrammetric performance results (after Leberl, 1983)

Source	Year	Accuracy (m) Along 1σ	Accuracy (m) Across 1σ	Relief	Control per 100 km²	Resolution (ground-meters)	Antenna Stabilized	Antenna Type	Radar system code	Scale of images 1:	Area designation remarks
Gracie	1970	20	14	Flat	10.0	17	Yes	Synth.	AN-APQ 102		Atlanta, Georgia
Leberl	1971	50	23	Flat	10.0	30	No	Real	EMI (UK)	200 000	Netherlands
Bosman	1971	47	60	Flat	10.0	30	No	Real	EMI (UK)	250 000	Netherlands
Konecny	1972	152	255	Mountains		17	Yes	Real	Westinghouse	216 000	N Guinea – conformable transformation
Greve	1974		35			≤3	Yes	Synth.	TOPO II	100 000	Digitized monoplotting
Good-year	1974	38	30	Flat	3.0	12	Yes	Synth.	GEMS 1000	400 000	Phoenix Arizona
Derenyi	1974	89	111	Hills	1.1	17	Yes	Real	Westinghouse	250 000	Washington, DC
DBA-Systems	1974	51	26		0.5	3	Yes	Synth.	AN-ASQ 142	100 000	Radar inferferometer
Konecny	1975	80	79	Flat		12	Yes	Synth.	GEMS 1000	400 000	Phoenix – conformable transformation
Derenyi	1975	30	28	Flat		12	Yes	Synth.	GEMS 1000	400 000	Phoenix – control density not specified
Tiernan	1976	209	257	Flat	0.7	30–150	No	Synth.	Apollo 17	1 Million	Lunar satellite
Leberl	1976	147	233	Flat	0.3	30–150	No	Synth.	Apollo 17	1 Million	Lunar satellite
Hirsch	1976	120		Flat	3.0	30	No	Real	EMI (UK)	100 000	Netherlands
Leberl	1976	140	190	Flat	0.3	25–150	No	Synth.	JPL L-band	500 000	Alaskan tundra – satellite radar simultaneous

showed that about twice the resolution length of the radar was the limiting accuracy. Work has also been carried out on the evaluation of radar for mapping purposes (Leberl, 1978, 1979, 1980 and Dowman and Morris, 1982) in particular Dowman and Morris examined the potential of radar imagery for identifying urban features compared with aerial photography and the results can be seen in Table 5.4. The study showed that the road network could be successfully plotted from radar even to being able to define single carriage roads as compared to dual carriageway roads. The conclusion was that SAR has a potential for small scale mapping and image revision but that more work was required to prove the potential.

During the Nigerian radar survey the radar strips were mosaiced to fit to the 1:250 000 Joint Operation Graphics map series, the climatic conditions for Nigeria are such that air photographs were available for the north but not for the south, or at best not as a regular survey acquisition. This meant that about two thirds of the country was mapped by photogrammetric methods and a high degree of correlation was possible between maps and images. For the remaining third in the south the maps had been constructed from traditional plane table field methods. While adequate control points could be identified to construct the mosaics, detailed comparison showed that although the general route and overall curvature of a river were compatible, the radar showed the river in much greater detail with more minor curves and bends than the generalized map.

Table 5.4 Type of urban building (after Dowman and Morris, 1982)

	Houses	High rise flat blocks	Housing rows	School	Works	Store	Church	Other
Number of units on aerial photography plot	62	54	16	23	17	3	10	4
Number of units on HH radar plot	23	38	5	17	15	3	3	3
Number of units on HV radar plot	18	47	3	11	10	3	1	2
On HH but not HV plot	13	1	4	6	5	0	2	1
On HV but not HH plot	9	11	1	0	1	0	0	1
% on HH plot	37.09	70.37	31.35	73.91	88.23	100.00	30.00	75.00
% on HV plot	20.96	87.03	18.75	47.82	58.82	100.00	10.00	50.00

Under the circumstances it was possible to revise the base map from the radar with the knowledge that although the result may not have been planimetrically perfect, it was a considerable improvement upon the existing map representation.

The ability of radar has perhaps to be considered in this aspect, in a highly developed country with a well developed mapping system such as the UK Ordnance Survey maps; it may be many years before radar can be considered as an alternative mapping system for planimetric or topographic maps. Most resources interpretation is however required for development studies and in many cases the map bases may not yet be completed to the Ordnance Survey type standards. Maps may be many years out of date or areas may be difficult to map and remain unmapped. Against this background radar can be seen to have a definite potential, especially from mosaics, to provide a reliable surrogate map base. There are circumstances where further survey work is dependent upon a map base as a more reliable position fixing aid. Radar mosaics can provide a useful navigational base for such surveys which could for example include airborne geophysical surveys of an area.

Where sound maps do exist, research work is being conducted to try to improve the geometric rectification of radar imagery by relating the imagery to a digital terrain model of the area; these studies have the potential to minimize some of the distortions described in this text, especially the foreshortening effect in mountainous country.

In countries where reliable mapping is available the generally accepted technique for mapping most resources is to use the imagery for interpretation and to transfer the results to a base map. To do this it is clearly necessary to be able to relate the image to the map with sufficient confidence to be able to transfer the data. In many instances precise fit is not essential since the data will relate to observable features, thus in crop studies the radar may be used to identify a crop unit but the unit will in turn be related to a field whose boundaries will mostly be clearly identified on the existing map and therefore accepted. Where changes occur there is frequently sufficient surrounding data to enable local correlation of the image to the map in order to compile the change.

The constraints for basic planimetric and topographic mapping from radar are not therefore a constraint to its use for most resources studies.

REFERENCES

Dowman, I. J. and Morris, A. H. (1982) The use of synthetic aperture radar for mapping. *Photogram. Record,* 10 (60) pp. 687–96.

Leberl, F. (1976) Imaging radar application to mapping and charting. *Photogram-metria,* 32 (3) 75–100.

Leberl, F. (1978) Current status and perspectives of active microwave imaging for geoscience application. *ITC Journal*, pp. 167–90.

Leberl, F. (1979) Accuracy analysis of stereo side-looking radar. *Photogram. Eng. and Remote Sensing*, **45** (8) pp. 1521–69.

Leberl, F. (1980) *Preliminary radargrammetric assessment of Seasat-SAR images.* Mitteilungen der Geodaetischen Institute No. 33, Tech University A-8010 Graz, pp. 59–80.

Leberl, F. (1983) *Photogrammetric Aspects of Remote Sensing with Imaging Radar – Remote Sensing Reviews*, Vol I, part 1, Harwood Academic Publishers.

Schanda, E. (1984) *A radargrammetry experiment in a mountain region.* SAR-580 Investigator's Final Report.

Ulaby, F. T., Moore, R. K. and Fung, A. K. (1982) *Microwave Remote Sensing Active and Passive*, Addison-Wesley Pub. Co. Inc., New York.

6

Project planning and field studies

6.1 INTRODUCTION

It can be inferred from the foregoing chapters that any project intending to acquire radar imagery will require careful planning, irrespective of whether the survey is to be conducted by airborne platforms or by spacecraft. This is, of course, equally valid for other imaging acquisition systems and some of the problems involved are common to both types of operations. Radar does, however, introduce its own peculiarities and these have to be given special consideration if the project is to be a success.

Of no less importance is the planning of the interpretation phase, which comprises the preliminary studies, the actual interpretation and the field checking. In this chapter attention is drawn to some of the needs and requirements for various aspects of the proposed project. Some requirements may appear to be obvious, some peculiar to a type of study and others may prove to have been unnecessary, the objective is not to provide here a check list but rather to draw attention to the overall need to give due consideration to all of the project in advance in order to achieve the maximum benefit and success from the survey.

6.2 JUSTIFICATION FOR THE SURVEY

Projects can be broadly divided into two major types, in the first the justification for the study is the need to carry out research into the applications of a technique to various survey requirements. In this case the planning of the project will be conditioned by the amount of funding available, it is impossible to accurately relate the costs involved to the benefits derived, indeed at the end of the project it may well be concluded that no benefit was derived.

Research projects vary from the small detailed studies carried out by manufacturing companies as part of their research and development programme or by university research departments, to the multi-national projects such as the European airborne SAR-580 campaign, Seasat, SIR-A or the two ERS-1 programmes.

The second type of project is that undertaken for a definite practical application: the state of the art operational programme. In such projects, available operational imaging radars are used in order to undertake a well defined survey for a particular purpose. Funds may be no less limited than in the research applications but in this instance a definite cost benefit is required, in other words one would expect to achieve positive practical results from the survey which would attain the pre-defined goals.

The justification for research is clear, it is essential to develop both instrumentation and data handling techniques, including interpretation, in order to provide a reliable operational system for use in practical applications. In the first instance it is a case of 'can this technique have any practical potential?' If the answer is yes as in the case of SAR, then further research is usually a result of experience derived from practical applications being fed back as requirements for the next generation of instrumentation. The development of the air camera into the current high precision instrumentation can be seen to be a result of such dialogues over many years and is still continuing.

In resources surveys, the justification for the use of SAR, whether airborne or satellite, will vary with the discipline concerned. Clearly the justification for a satellite system can rarely be made by any one agency alone and is normally considered as a major national or international project, the results of which will be both research oriented and of potential practical benefit. In the early stages the use of satellite SARs cannot be justified on the basis of practical resources benefits but must be considered as a research development although clearly long-term practical benefits are intended.

Thus we are left with the need to justify an airborne survey for a resources study. In this case funding may be by a private corporation, for example an oil company, by a national body such as a Ministry of Agriculture, or by an international loan agency providing funds to a national body.

The decision to use a SLAR for a survey may be made on the basis that climatic conditions preclude the use of a visible band system and the all weather capability of the radar makes it the only alternative. If this is the case then it is essential to ensure that the instrumentation used and the state of development of the technology will indeed produce imagery which is capable of the levels of interpretation necessary for the proposed survey.

This may seem obvious but it has to be remembered that published favourable research results produced in a laboratory environment may not be equally reliably reproduced under operational conditions over a wider area. The research may have been carried out over a limited area with clearly defined targets, under ideal conditions with no restraints on field measurements, conditions which it would be impossible to reproduce over a wider target area or in areas of limited or restricted access.

Because cereal crops can be reliably interpreted in say Europe it does not necessarily follow that rice can be equally reliably identified in Asia, similarly tree crops in these two environments will be substantially different and the positive results obtained in one region may not be replicated elsewhere. This is not to suggest that SLARs cannot be used in a variety of operational surveys, they can and have been most successfully used, rather it is necessary to point the need in using SLARs to ensure that the potential of a system is fully understood and the anticipated results clearly related to the potential of the system used.

The main justification for using a SLAR is usually given as its all weather capability. Clearly this is a major attraction particularly in those regions where persistent cloud cover makes the regular and reliable acquisition of visible imagery almost impossible. SLARs do have a potential beyond this all weather approach however, and this must not be overlooked.

It has been shown that the use of a SLAR can enhance the expression of relief and that in certain areas, such as tropical forest areas, the effect of tree cover can be suppressed when using certain wavebands, to the benefit of the expression of relief. This potential to be able to see physical structures is of great interest to geologists and therefore SLARs have the potential to give valuable interpretative data to a geologist which is equally valid whether the area is obscured by cloud or not.

For a geological survey, therefore, the justification for using a SLAR survey would be on the basis of the image derived from using SLAR rather than its all weather capability alone.

There is growing evidence that the use of multi-images in combination, whether mixtures of look directions, frequencies or polarization or all three, can yield improved data to enable crops and vegetation units to be discriminated with increasing reliability. If this proves to be the case, then SLARs, particularly SARs, may have a future for land use surveys which, like geology, is beyond the all weather capability.

6.3 FLYING OPERATIONS

All survey flying operations require careful, detailed planning, SLARs however introduce additional requirements that have to be taken into consideration.

In discussing radar shadow it was shown how SLARs can express a directional bias and this bias can lead to misinterpretation of data especially for geology. The layout of the flight plan has therefore to take into consideration the general geomorphology, or grain of the country, in order to minimize the effect of bias and to use radar shadowing to maximum interpretive advantage. Such a plan does assume that the existing data for an area are sufficiently accurate to provide this information.

As with all flying operations all the usual clearances are necessary from civil and military offices, these may well have to be extended in some circumstances if the ability to image at night with SLARs is used. This potential may be particularly useful in areas where there is considerable day time air traffic, such as major flight routes near a principal airport. The ability to image during non-daylight hours may also be used in order to obtain maximum operation of a SLAR within a desired time span.

In some areas night operations can impose new constraints, some authorities restrict night operations because of potential noise pollution and in some instances vital support services only operate during daylight hours.

Much is made of the ability to obtain imagery using SLARs in all weathers except during periods of excessive precipitation. This is a major factor in satellite operations where no serious constraints are experienced by the platform, in airborne operations, however, it depends upon the aircraft platform itself being able to operate.

In the first instance the aircraft has to be able to become airborne, this may seem obvious but fog and mist which, for example, may not affect the operation of a SLAR may well close an airport. This may require careful planning to avoid operating from airports where such constraints are known to occur or in some areas choosing the right season to avoid physical phenomena which can constrain flying, for example dust storms or high winds such as the Mistral.

In visible imaging systems the traditional method of planning survey flights is to lay out the flight lines on a map and to fly using visible navigation to obtain final positional/directional accuracy, although most surveys now use precision radio beacon navigation. To ensure complete area coverage, rush prints obtained by initial flights are used for subsequent navigation. The all weather potential of SLARs would result in a high probability of flying during extreme cloud cover conditions, thus any form of visible reference navigation cannot be undertaken.

The aircraft has to be fitted with the most advanced navigational aids such as inertial systems, satellite referencing and so on, if it is to operate successfully. Fortunately this is not the constraint it used to be as the global network for precision position fixing improves through the use of satellite systems.

In earlier chapters where the radar geometry was being considered the method of correcting for the oblique view of radars was described. It will be recalled that if ground range correction is required the height of the ground surface and the platform have to be accurately recorded. From this it follows that any aircraft used has to be equipped with a precision altimeter giving a continuous flight record and that the relative underlying terrain height is necessary. Clearly this condition can only be met in those areas of the world where there is a reliable contoured map base.

Aircraft attitude has also to be reliably measured and recorded, in most systems the data are fed directly into the radar instrumentation and used to perform an immediate data adjustment. It will follow from this that excessive changes in platform attitude will seriously affect the subsequent imagery. In certain areas of the world climatic conditions are such that periods of excessive air movement or turbulence can occur. This is particularly true over the more arid regions where seasonal dust storms occur such as the Harmattan in northern Nigeria. The effect of turbulence has to be recognized and flight planning needs to take any known seasonal occurrences into consideration in the same way as in planning photographic surveys.

As a final consideration it is necessary to understand that radar is an active microwave system using a high frequency energy pulse generated on the platform used. This energy can have an effect upon certain ground systems, notably any ground radar receivers that may be oriented towards the imaging aircraft. The passage of the aircraft will mean that the exposure to the radar beam is for a very brief period, but the sudden energy surge in a radar receiver could produce more serious short-term errors. It has occurred during some flying operations that major defence radars have been blacked out by the radar beam, causing system shut downs in the defence radars for some minutes. Any sites within flying operations areas should therefore be warned of the potential of this occurring and adequate clearances obtained.

6.4 DATA COLLECTION

It is an essential operation in planning any resources project to accumulate and to study all relevant data on a region during the planning stages and throughout the study. There is a tendency, particularly in remote sensing, to ignore the potential of other data sources and to carry out interpretation using only the imagery obtained. Figure 6.1 is a block chart of a typical radar survey, it could however apply to any similar image based survey.

It will be seen that the first task was to obtain every available existing data source whether they be maps, reports, statistics or other imagery bases, and to examine these in order to become acquainted with the area. In this instance aerial photography was available at a variety of scales and of widely

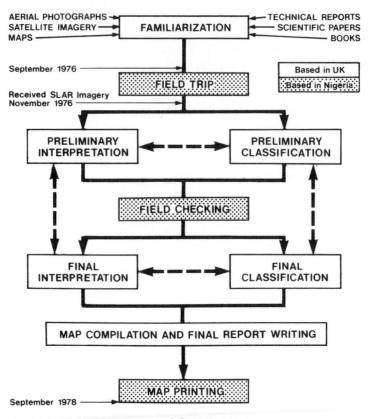

Fig. 6.1 Airborne survey operational flow diagram.

variable dates. In most cases the photography, however old, still yields valuable information about an area, the vegetation types, settlement patterns and so on. Throughout the interpretation recourse is made to these data bases in order to clarify points of doubt or to help with a problem in interpretation.

A study of the available data sources will enable survey flights to be planned to best advantage and identify areas where supplementary field studies may be required.

The diagram shows that pre-interpretation field studies have been allowed for, these are desirable but not always possible, indeed in the more remote areas the very inaccessibility may be the reason for the survey and thus by definition field studies may be impractical.

The type of pre-interpretation field work will vary with the type of survey, the plan shown in the figure referred to the Nigerian radar survey covering the whole country. In this case there was a large amount of regional homogeneity and well planned field trips helped the interpreters to become acquainted with the major interpretation units that could be expected.

In other areas the level of final survey may be much more detailed and the need for a more comprehensive field study increases. If the date and time of survey overflights can be determined in advance then some field studies at the time of overflight should be undertaken.

During the European airborne radar campaign (the SAR-580 campaign) it was clear that many of the investigators had no experience of radar and that since the results were to be used for research studies, field data were essential. To assist the researchers a team of specialists or discipline leaders met to establish the fundamental requirements for field data collection. These were subsequently compiled into a campaign handbook for investigators to follow.

The requirements varied with the particular discipline concerned but in the main the rules were to observe and note everything in an area, even items that may have been considered irrelevant to the discipline involved. For example, note should be made of potential major radar reflectors, these may be large water towers, power pylons or existing radar receivers, items which could cause a high signal return to give a point spread over an area many times larger than the actual target and cause interpretation confusion. Climatic conditions could be significant, if a flight plan is scheduled to last over a number of weeks, variations in climate may be occurring which could affect the image appearance. It has already been shown that SLARs are affected by the dielectric constant of targets, this means that changes in the available water, which could influence this constant, could alter the backscatter response of an otherwise identical target. Images may well span a dry spell and run into a wet season giving rise to changes in image appearance over an area. The all weather capability of SLARs can easily lead to the assumption that the weather conditions were constant.

The backscatter response can be a factor of the leaf size and angle, this will in turn vary between a still air day and a very windy day. The effect of the wind on leaves can change the target signature on an image, furthermore the wind over a dense standing crop such as cereals can be seen to be comparable to water wave motions, this in turn can give the Rayleigh effect.

At this stage in the development of radar techniques it is important to recognize that much less is known about the backscatter returns of targets and their eventual image expression, than is known about optical imagery; at the same time the optical imagery is more closely related to human visual experience. This being the case it is necessary to be more observant of objects and phenomena in an imaging or survey area and to be able to recognize the possible effect on an image. Even so it is not possible to pre-examine all potential influences on an image, for example the expression of sub-surface river systems in desert areas observed on shuttle imaging radar

SIR-A was not foreseen and came as a surprise to interpreters. In a similar manner Seasat imagery of coastal areas provided clear evidence that the sub-surface topography could affect the wave surface to such an extent that the image reflected those bathymetric changes.

Figures 6.2 and 6.3 are two field sheets used for the Nigerian survey. They serve to demonstrate the range of data collected during field surveys to support and assist the eventual unit classification. The survey was intended to map vegetation but data were collected on soil structures, geology and geomorphology in order to build up a useful data profile of a typical interpretation unit.

Where detailed topographic maps already exist it may be found useful to mark all information onto a key base map and Fig. 6.4 shows how a map used in the SAR-580 study referred to earlier was compiled.

In all forms of imaging no system has yet been devised which can accurately and consistently identify mapping units. The confidence level of interpretation may be exceedingly high in some instances based upon existing knowledge of an area and accurate correlation with spectral signatures for optical imagery. Much will depend upon the experience of the interpreter and the more experienced the interpreter the more they will recognize the need for reliable field checking.

6.5   INTERPRETATION

Specific aspects of interpretation will be addressed in subsequent chapters, at this stage it is regarded as desirable to establish the processes or estimators used in interpretation.

It may seem an unnecessary exercise to start at basics in interpretation at this stage, one could assume that most students of the potential of radar images will be experienced at interpreting conventional visual imagery. This however is exactly the problem, we are used to identifying objects on a photograph because they relate to human experience. Our eyes see in a photographic sense therefore a photographic record is our normal view of the world. Multispectral combinations with strange colours can introduce a new dimension but even here the underlying structure is immediately recognizable.

The radar image is not a normal view, and whilst many objects and shapes may be instantly recognizable, for example geological structures, in many cases a definite interpretation rather than recognition has to be made. The objective here is therefore to re-establish the basics of interpretation and to try and anticipate potential types of interpretative error. The processes of interpretation are usually divided into three main headings namely, tone, texture and contexture.

VEGETATION RECORD:

LOCATION: OBSERVER: DATE:

Observation of region

1. PATTERN   AREA UNDER REFERENCE

Indicate regional vegetation   PATTERN SKETCHED
pattern on chart and describe: SCALE

		Scale	1 cm	1 cm²
MATRIX			1:10k	100 m	1 ha
		NONE	1:100k	1 km	1 km²
			1:250k	2.5 km	6.25km²
ELEMENTS WITHIN		RANDOM			
		REGULAR			
		CLUSTERED			
		LINEAR			
		OTHER			

Indicate whether matrix or element at local site

Observation of local site

PHYSIOGNOMIC TYPE   CHARTERS VEGETATION TYPE

DISTURBANCE CLASS

CHARACTERISTIC DOMINANTS

	EMERG.	MAIN	LOWER	GROUND
HEIGHT CLASS				
COVER PER CENT				
CANOPY TYPE				
CANOPY DENSITY				

PROFILE   Height Scale

PHOTOGRAPHIC REFERENCE

Fig. 6.2 Vegetation record field sheet specimen.

TERRAIN CLASSIFICATION — REGIONAL DESCRIPTION

SAMPLE AREA NO: OBSERVER: DATE:

LOCATION:

MAP REF: SLAR REF: PHOTO REF:

LANDSAT REF: ELEVATION: RAINFALL:

LAND USE: EXTENT OF AREA UNDER REF:

GEOLOGY	
LANDFORM	
TOPOGRAPHY	
DISSECTION	
SOILS	

SITE DESCRIPTION

SLOPE: MICROTOPOGRAPHY:

EVIDENCE OF EROSION: NATURE OF SURFACE:

DRAINAGE:
        SURFACE: SUB-SURFACE:

SOIL
DESCRIPTION:
                BORE [ ]   PIT [ ]   SECTION [ ]

HORIZON	DEPTH CM	COLOUR	MOTTLES	CONSIST.	TEXTURE	OTHER PROPERTIES

Fig. 6.3 Terrain classification field sheet specimen.

Fig. 6.4 Sample field data map for vegetation survey. Note how observations include objects that could affect the radar backscatter such as metal covers, electricity transformers and construction work as well as notes on row direction, boundary type and crop state.

6.5.1 Tone

In a single image, tone is taken as the overall grey level of a unit, units may be differentiated from each other on the basis of these tonal differences or may be considered to be comparable because of tonal similarities. This cannot be established as an overall rule however as tone can vary with aspect, the same unit on a slope facing the platform will probably display a different tone to the same unit on the hill side facing away from the platform. Water bodies usually appear as black shapes on an image, radar shadow is also black and in some cases can produce a shape that can easily be misidentified as a water body and vice versa. Thus tone can be an important estimator and in some cases may be the only criterion on which two units are separated, it is not however infallible. If colour images through combination of scenes are used then the colour differences themselves become the determining estimator with sub-divisions of colour arising from differences in the tones.

Most of the smoothing routines that have been developed are aimed at providing an image of varying tones so that any discrimination is carried out mainly on the basis of tonal variations. This would be satisfactory if tone were the only estimator used by an interpreter, but this is not the case, most interpreters also rely on the next estimator as well – texture.

6.5.2   Texture

There is considerable discussion between skilled image interpreters and physicists involved in image construction and analysis as to whether all the ranges of texture categorized by the interpreter can be said to exist in digital analytical terms. In any automated analysis it would be essential for a computer to be able to recognize a textural pattern and to quantify it in such a manner as to be able to identify other units of comparable texture, so far the existence of all the variations in texture cannot be reliably recognized in this context, although the more coarse textures can be seen to give rise to a measurable difference.

In the case of radar the texture of an image is in part due to the speckle on the image. This speckle is a system induced artifact and system development is aimed at reducing the amount of speckle in an image. To most interpreters the existence of speckle texture is obvious and very real and is frequently one of the main interpretive criteria used, especially in forestry studies.

The image in Fig. 6.5 is a piece of SAR imagery of a forest plantation area surrounded by agricultural fields. An interpreter would recognize between five and six different textures ranging from very smooth to very rough. A grass field would tend to exhibit a very smooth texture whilst mixed woodland becomes very rough in texture. Figs 6.6 and 6.7 are enlarged segments of the image designed to demonstrate the variations of texture. In the case of a very smooth image the presence of texture is almost non-existent, some internal variations of tone do occur but in the main the scene can be said to have a very even tone. As one moves through the textural ranges the tonal mix within a unit increases, initially a scattered variation occurs but gradually the units of tone group together to form a pattern of larger but random blocks of tone.

The basis of radar imagery is to measure the backscatter of a target and the changes in backscatter are seen to be a factor of surface roughness, in woodland there is a roughness of each tree and a secondary variation caused by the variations in heights between trees giving an additional roughness.

Fig. 6.5 SAR imagery of a forest area.

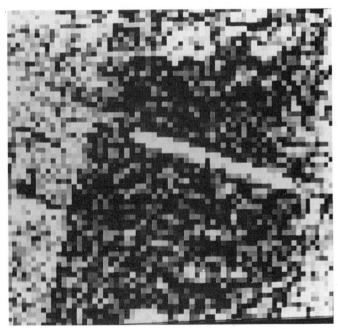

Fig. 6.6 Part of Fig. 6.5 enlarged to demonstrate very smooth texture.

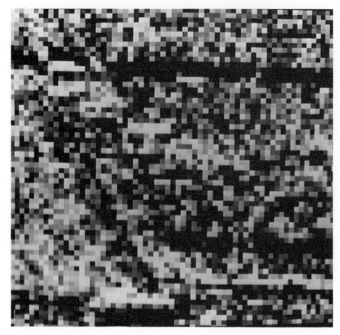

Fig. 6.7 Another section of Fig. 6.5 enlarged to demonstrate very rough texture.

There is therefore a natural roughness of signal return occasioned by the target itself and a speckle induced by the system. The problem with speckle reduction therefore comes in determining the extent to which what is being removed can be said to be due to the system and in ensuring that roughness due to the target itself is left for the interpreter to recognize as true texture.

6.5.3 Contexture

The degree to which units can be discriminated may be largely based upon tone and/or texture. The potential identification or interpretation of these units may require recognition of other factors and these are variously grouped under the heading contexture.
Contexture can be taken as:

Size – The size of a unit may well help to provide an indication of its identity; as a demonstration, man made objects will usually be limited in size whereas natural objects will not be so limited, this is best illustrated by natural forests as against plantation.
Shape – The shape of an object is usually the estimator most related to human experience, the interpreter recognizes a continuous meandering line of varying width as a river, or a rectangle as a field and so on.

Association – This is taken as the relationship of a unit with other recognizable interpretative factors, thus in a tropical area wooded areas following a river are associated with the river and are interpreted as riverine forests, a target area may only be recognized as an irregular, rough, light toned patch but the presence of radial lines and a surrounding network of fields could lead to the conclusion that the object is a habitation area or town with the radial lines being roads.

Human experience – Most important of all is the experience of the interpreter, in their own discipline and in the field identification of objects compared to their radar signature.

It has been described earlier how in geology the radar image has been shown to be of great benefit because of the manner in which structures are enhanced and presented. The value of the image is however dependent upon the ability of the interpreter to recognize the structures and to derive an understanding of the geology of the area from the data presented.

In other disciplines the interpretative evidence may not be so dramatic and so evident; as we have seen, differences of tone and texture may provide discriminants and inferences made from other contextural data. Finally however the interpreter has to try to place the interpreted units into context and this requires the application of their special knowledge and experience.

In examining optical imagery recognition is comparatively easy but in radar imagery, because a backscattered energy signal is not the normal view, recognition may not always be obvious. Examination of some of the examples of radar in this book would, at first glance, seem to contradict this statement. In many cases objects are easily recognizable, field patterns, roads, rivers and mountains are evident. Closer inspection usually reveals anomalies or deficiencies, in one area a unit may be identified whereas in another scene a similar unit may not be so evident, in town areas the bright signals dominate the scene and roads cannot be seen. It is for this reason that the interpreter has to return to the basic steps of interpretation and then have the ability to read the data so derived.

A simple example of human experience is given by Fig. 6.8(a) which most people would recognize as a house. Analysis will show that neither its tone nor texture are especial factors in identifying the house; there are shape and size, but then houses vary in shape and size and still remain identifiable; the main factor is human experience; people have seen houses and therefore recognize houses readily. In the next figure (b) the elements of the house have been reduced to a few basic lines, nevertheless most people would still be able to see a house. So human experience may identify the target, and even when only partial interpretive evidence is available it is still possible to conclude from this evidence that a house is the target present. In (c) we introduce another element, the house can be deduced but the elements of the

Fig. 6.8 Human experience applied to interpretation.

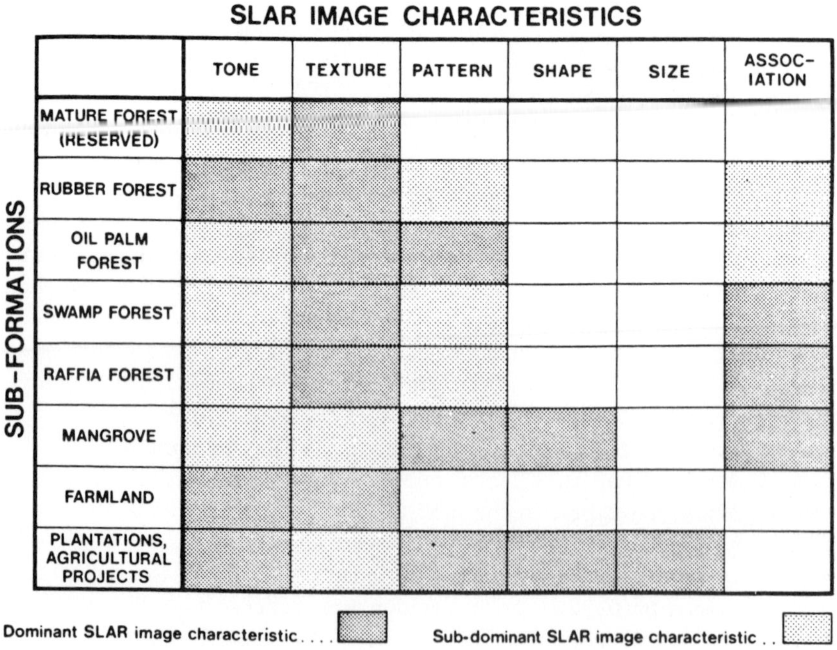

SLAR IMAGE CHARACTERISTICS

SUB-FORMATIONS	TONE	TEXTURE	PATTERN	SHAPE	SIZE	ASSOC-IATION
MATURE FOREST (RESERVED)	░	░				
RUBBER FOREST	▓	▓	░			░
OIL PALM FOREST	▓	▓	░			░
SWAMP FOREST	▓	▓				▓
RAFFIA FOREST	▓	▓	░			▓
MANGROVE	▓	▓	▓	▓		▓
FARMLAND	▓	▓				
PLANTATIONS, AGRICULTURAL PROJECTS	▓	░	▓	▓	▓	

Dominant SLAR image characteristic . . . . ▢   Sub-dominant SLAR image characteristic . . ▢

Fig. 6.9 SLAR image characteristics.

Fig. 6.10(a)

Fig. 6.10(b)   *caption overleaf*

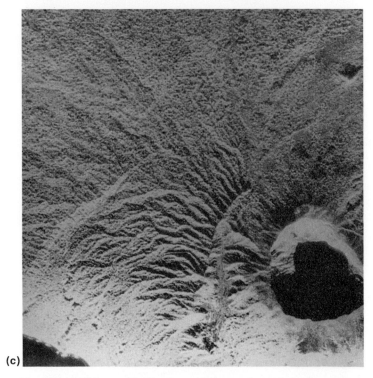

(c)

Fig. 6.10   The need for a wide data base to accurately apply human experience. (a) Given only a segment of the image the interpreter may conclude that a fairly flat forest area is being observed with possibly some form of delta pattern. (b) In the larger view relief is apparent and it could be deduced that this is a ridge or a gentle flat slope, but the direction of flow of the drainage is perhaps uncertain. (c) In the full scene the picture is clear, a volcano with an inner and outer crater and very steeply sloping sides.

tree are substantially different, an interpreter should be able to conclude that the images were obtained at different seasons. In (d) we have another target which not everyone would so readily recognize, the target may be recognized as a house again but the total expression is completely different from the first house, there are very few comparable factors. The identification of the target relies on the probability that the interpreter has had field experience in a region where such houses, or ones similar, are usually found, on the other hand they may have seen pictures of similar houses thus using an ancillary data source to aid interpretation.

Figure 6.9 is an interpretation chart produced for a vegetation study of radar in Nigeria. The chart shows the factors which predominate in the interpretation of various units. Natural forest has a distinctive tone but texture is the main discriminant, the contextural units have no role to play since neither the size of a forest, nor the shape of a forest are pre-definable nor is there any predictable association with other features. With swamp forest the situation is different, tone and texture still have a role to play, but on these alone it would just be forest, the association with rivers and the detail pattern all help to classify the unit as swamp forest. The other units main criteria can be seen from the chart.

Figure 6.10 is an example of how human experience can make the wrong conclusions.

Finally, as with optical multispectral scanning, colour composites will result in colour combinations that are not within human experience, interpretation will rely on the ability to recognize units and to relate them to human experience. It should be borne in mind that not all persons have the same ability to identify or discriminate colours, ideally interpreters should be checked for colour perception before using digital image analysis displays.

7

Airborne systems

7.1 INTRODUCTION

Airborne systems can be broadly divided into three groups

(a) the less complex, real time SLARs;
(b) the main commercial operation systems; and
(c) the advanced experimental systems.

In the examples quoted, with one exception, only the systems known to be in operation are referred to and only then those systems that are non-military. There have been a number of systems built for restricted experimental use but it is unlikely that resources interpreters will ever come across any image products for evaluation. In the one system, the Westinghouse, that has been included, this was a commercial system and its image products of large areas may still yield valuable data for certain studies. Table 7.1 shows some of the main imaging radar systems.

7.2 REAL TIME SLARS

In their simplest form some of the SLARs are not imaging radars in the sense that a full pictorial display results, in many cases the screened image will resemble the radar screen images normally associated with on-board ship radars or aircraft monitoring systems. Exceptional targets such as ships and shore-lines are revealed as bright responses against a dark background. The Ericsson SLAR is an example of such a unit and consists of an antenna, signal processor, control panel, transmitter receiver and a TV display. It is readily mounted on almost any light aircraft and is an X-band (9.3 GHz) vertically polarized radar with 10 kW output power.

The radar provides a side scan, which is designed for maritime surveillance so that ships will appear on the video screen as bright spots. It can be configured to produce a dual look capability, that is it will scan both sides of the aircraft thereby widening the effective field of surveillance.

The basic system can be improved to provide a more recognizable image on the video display and the video display can be recorded (Fig. 7.1).

The Interdan system can be seen as an extension of SLAR systems such as the Ericsson, in this case the image is a recognizable image which can, with adjustment, be adapted to land application. It has a limited use partially due to its poor resolution – but it may well be ideal for monitoring hazards and phenomena such as floods or extent of snow. It is already in effective use monitoring oil spills in oceans. This particular unit can provide a simple, real time, hard copy or a video recording. In operational use for monitoring it can transmit a continuous real time video picture to a ground receiver so that phenomena can be observed by ground parties.

There are a number of other similar radars available, these two have been selected as typifying the systems.

7.3   COMMERCIAL OPERATIONAL RADARS

The three main commercial airborne radars of the sixties and seventies all derived from military radars and were originally developed to conform to research specifications set under US military contracts. Of the original three, only two remain. Table 7.2 shows some of the major surveys undertaken by these three systems and it is clear that large areas have been so covered. Figure 7.2 shows the extent of areas related to a global map.

The apparent high cost of the surveys and lack of understanding of their potential, particularly for developing countries probably accounted in part for their restricted continuance, coupled with the overall economic climate of the late seventies and early eighties and the overall lack of experience in interpreting the final data. In the early eighties a number of more intensive campaigns and the results of Seasat and Shuttle radars have renewed interest in the potential of systems and a wider circle of users are more aware of the data potential.

7.3.1   Westinghouse

As Table 7.1 shows this was a K-band SLAR of 25 m resolution with both HH and HV polarization potential. Under normal operational conditions it produced real time film imagery at a scale of 1:250 000 and was flown on a Douglas DC-6B. A large number of projects were undertaken using this

Table 7.1 Airborne imaging radar systems

Manufacturer	Model	Type	Wavelength (cm)	Nominal resolution			Polarization	Comments
				Along track (at 15 km) (m)	Across track (m)	Ground swath width (km)		
Westinghouse	APD-7	R	0.86					
	APD-8	R	0.86					
	APQ-56	R	0.86					High altitude system.
	APQ-97	R	0.86	25	8	21	HH and HV	Operated from DC 6B Ceased commercial operation 1973.
Goodyear	APD-10	Syn	3.1	10	10		HH	
	APQ-73	Syn	3.1					
	APQ-102	Syn	3.1	15	15	37		Ultimately the GEMS unit.
Texas Instruments	APQ-55							Currently mounted in a Caravelle aircraft.
Motorola	APQ-86	R	0.86	46	15	16	HH	
	APS-85	R	2.5	116	75	100	HH	
	APS-94	R	2.5	116	75	100	HH	
	APS-94D	R	2.5	116	30	100	HH	Mounted originally on a Grumman Mohawk, latterly on a Grumman Gulf Stream.

General Dynamics	APQ-69	R	3.1	380	15	25	H and V	
Philco–Ford	DPD-2	R/Syn	1.8	30–60	12–25	18	H and V	
Kelvin–Hughes	–	R	3.0					UK
EMI	P 391	R	0.86	52	15	28	HH	
University of Michigan	WW	Syn	3.1	10	10	18	H and V	Ultimately the SAR 580 system.
			3.4				HH	Mounted on a Convair 580.
RCA/NRL			6.7				HV	
			24.5				VV	
			70.0				VH	
Thomson CSF	RAFALE	Syn	3.3					
Thomson CSF	VIGIE	Syn	3.22	30	15	21	HH	Operated by GDTA (France). Mounted in a B17.
Ericsson		R	3.1	75	75	21 km+		Low resolution, real-time, surveillance system.
JPL/NASA	–	Syn	25.0	25	25	–	HH, HV	
Toros	–	R	–	–	–	–	–	USSR Dual-look system.

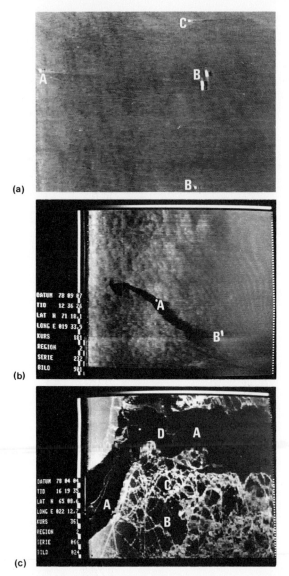

Fig. 7.1 Uses of a real time SLAR (courtesy of Ericsson). (a) Monitoring shipping. Naval ships detected at 3 km (A) and 12 km (B). The small radar echo (C) represents a partially submerged submarine. The wakes behind all ships are clearly displayed. Note also the radar shadows behind the larger ships. (b) Oil spill detection. Twenty-five tons of crude oil in the North Atlantic. The oil was dumped about 24 hours before mapping and has expanded to a slick approximately 10 km × 2 km. Two ships were present at A and B. Wind speed 3–4 ms^{-1}; light rain. (c) Sea ice surveillance. Sea ice in the Bay of Bothnia. Open water (A) normally gives less backscatter than ice, but can be confounded to level ice (B) if geometrical shapes are not considered. Ice ridges (C) give strong returns as well as drift ice (D).

Table 7.2 Summaries of SLAR system activities (up to 1980*)

Surveys by Westinghouse earth resources SLR mapping system

	Km²
Canada	91 000
Panama	20 800
Colombia	156 000
Ecuador	62 000
Nicaragua	6 500
United States	850 000
Australia	52 000
New Guinea	182 000
Indonesia	97 500
Nicaragua	128 000
British Solomon Islands	6 500
	1 652 300

(Ceased mapping activities in 1973)

Surveys by Aero Service Corporation

Brazil	8 500 000
Venezuela	900 000
Colombia	320 000
Peru	600 000
Appalachian area, US	100 000
Alaska	700 000
	11 120 000

The Caravelle is known also to have operated over the UK and in SE Asia

Surveys by Motorola Aerial Remote Sensing Inc.
(data to 1st July 1979)

SE Asia	40 000
Central America	71 000
USA (including Alaska)	1 678 000
Canada	80 500
Africa (five countries)	1 014 000
United Kingdom	3 000
	2 886500

* Note because of commercial confidence companies were reluctant to divulge data on surveys after this date.

radar as can be seen from the table and these images were successfully interpreted by geologists, foresters and land use specialists. Figure 2.19 is a typical Westinghouse image and it will be readily appreciated that the image demonstrates a good tonal range and a consistent resolution quality. Many interpreters regarded this instrument as producing the most valuable radar imagery for their purposes and reference is made here to the system because of the amount of useful coverage that still exists and could be of interest in certain future studies.

In 1973 the Westinghouse company decided that development and operation of an airborne radar was not compatible with their other company activities and did not have adequate market potential to warrant continuation as a special ancillary service, the facility was therefore withdrawn and is no longer available.

7.3.2 Motorola system

Motorola is also a real aperture system, an X-band radar with film recording. The system is a dual look imaging radar which has operated successfully for a number of years.

As in the case of Westinghouse, Motorola Incorporated, the original parent company, decided the operation of an airborne SLAR did not fit their product range and they discontinued operation and development. Unlike Westinghouse however, they sold off their aircraft and system to another company and allowed them to continue to trade as MARS, originally Motorola Aerial Remote Sensing Inc. This system is therefore still an operational radar.

Figure 7.3 shows the system fitted to a Grumman Gulfstream and Fig. 4.1 shows the real time product being produced for study by the operator. The use of a dual look facility and the type of aircraft helped to achieve economies in flying and the system is economically viable as an imaging source.

The Motorola radar is a modified APS/94D SLAR in the X-band 9315 MHz horizontal polarization radar using a 5 m antenna which can provide imagery at 1:250 000; 1:500 000 and 1:1 Mn scales using swath widths of 25 km, 50 km and 100 km at a 30 m resolution. The processing is real time with an inflight monobath development producing an original film display. Figure 7.4 shows the dual look capability on the operational system. Furthermore the image as it is acquired and processed in film form is almost immediately displayed to the operator ensuring continuous quality control. The operator is in continual contact with the aircrew and can therefore report immediately any deviations in course, warn when aircraft altitude is seriously affecting imaging and even cancel imaging if such faults become unacceptable.

Earlier chapters dealt with radar shadowing and described how detail is

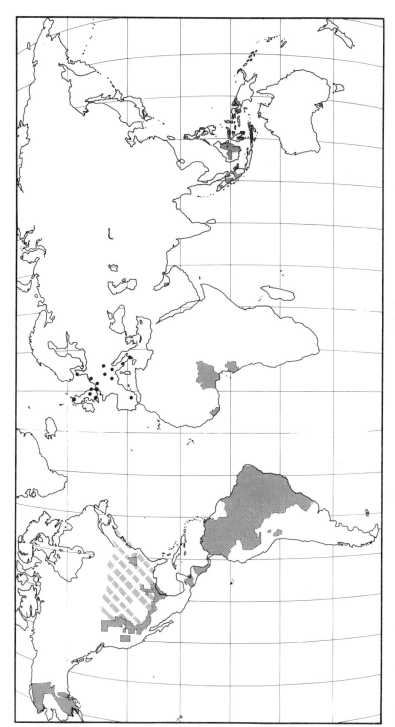

Fig. 7.2 Areas of known airborne radar data, including the SAR-580 areas in Europe and areas in the USA known to have been covered but for which no precise boundaries are available.

Fig. 7.3   MARS radar fitted to a Grumman G-159 aircraft (courtesy of MARS Inc.)

lost in shadowed areas, to overcome this it is usual to try to image an area from two look directions. The MARS system is a compatible transmit and receive antenna. This type of system is unique and it does provide economies of operation. Figure 7.5 is an example of MARS imagery.

When considering optical imagery the dual look presents very few problems, the standard of imagery is comparable and to the visual interpreter it would be difficult to recognize differences. Differences can and do occur however, in one known case a difference in imagery was found to be

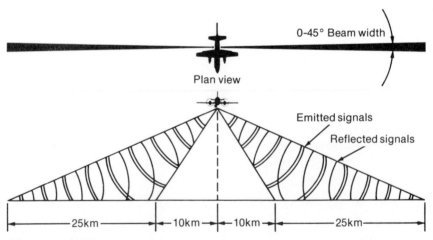

Fig. 7.4   MARS Inc. simultaneous dual look mapping mode for 1:2500000 scale imagery using 10 km delay. (Courtesy of MARS Inc.)

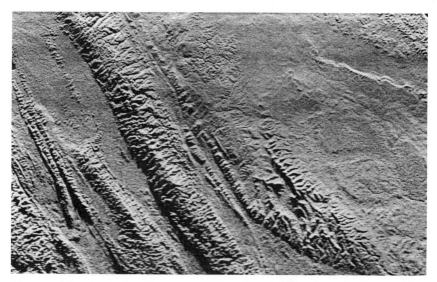

Fig. 7.5   An example of MARS Inc. imagery. This image was taken in April 1976 during a SLAR project of the Petén area in Guatemala.
Radar: Motorola modified APS-94D, X-band, real aperture, simultaneous dual-look antenna.
Scale: 1 : 250 000
Altitude: 13 000ASL
Look direction: south
Delay: 10 km
Time: 10.00 am
Weather: 80% cloud cover.
By Motorola Aerial Remote Sensing, Inc., 4350 E Camelback Road, Phoenix, Arizona, 85018.

attributable to water in one of the antenna pods which entered in a particularly heavy rain storm.

With improvements in the analysis of radar imagery through the use of computer aided analysis there will be a greater dependence upon total system calibration and on precision digital recording and processing. In these circumstances the use of dual look could introduce problems, for example it would be essential that both antennae have compatible characteristics.

The MARS system also offers control of the depression angle; this is usually used to achieve the range of imaging scales or to vary flying heights with variations in terrain but yet maintain imaging scale. The variable depression angle could be significant in certain geological studies by helping to emphasize microrelief which could be geologically significant. Table 7.3

Table 7.3   Depression angles at different altitudes (from Mars Inc.)

	25 km swath width '0' delay		
Altitude (ft)	Near R (degrees)	Far R (degrees)	Mid-R SW used (degrees)
20 000 (6098 m)	45	12	29
12 000 (3658 m)	36	7	20
8 000 (2439 m)	26	5	14

	25 km swathwidth '10 km' delay		
Altitude (ft)	Near R (degrees)	Far R (degrees)	Mid-R SW used (degrees)
20 000 (6098 m)	31	10	22
12 000 (3658 m)	20	6	14
8 000 (2434 m)	14	4	9

shows some of the variations in depression angle possible with the MARS radar while Fig. 7.6 relates the depression angle to the ground swath and/or the flying height. The selection of the correct depression angle is an essential element in pre-operational flight planning.

The real time recording/display is achieved by displaying the processed signal on a Cathode Ray Tube (CRT) display as a scan line, which in turn is focused by a camera lens onto a section of 24 cm wide photographic film. The film passes over the CRT at a rate commensurate with the speed of flight so that subsequent pulses at 750 per second are recorded. The film is instantly wet processed and then passed immediately over a light box so that it can be viewed by the operator (Section 4.3).

The MARS aircraft has the facilities for carrying other sensors, for example it is possible to install a precision aerial camera to obtain aerial photography. It should be remembered that although the imaging devices may operate simultaneously, the images will differ since the radar is side looking and the camera will observe vertically directly below the aircraft.

The MARS system has been, and is still, used extensively in the USA and throughout the world with considerable success for a number of resources surveys. The system was the one used for the National Survey of Nigeria and for Togoland. The Nigerian work is described in greater detail in Chapter 9.

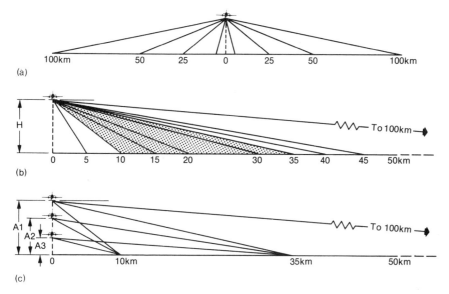

Fig. 7.6 Depression angle related to ground swath width. (From MARS Inc.) Diagrammatic sketch showing (a) maximum swath width and selectable 25 km and 50 km swath width options, (b) selectable 25 km swath widths using delays, and (c) varying depression angles by changing flight altitude without changing swath width or mapping scale.

7.3.3 Aero Services SAR System

This was formerly known as the Goodyear system and unlike the Westinghouse and MARS systems, it is a synthetic aperture radar. The system is known as the Goodyear Electronic Mapping System or GEMS radar. Like MARS it is an X-band radar with the following specifications:

Radar:	Frequency	9.6 GHz
	Bandwidth	15 MHz
	Dispersion ratio	15:1
	Transmitted power	50 kW peak
	Receiver noise figure	5 dB
	Transmitted pulse	1 μs
	Recorder pulse width	0.06 μs
Resolution:	Azimuth 10 m	
	Range 12 m	
Recorder data:	37 km of imagery in two channels ground range presentation	
Aircraft:	Caravelle Twinjet	

Since the system is a SAR it is not possible to generate real time film imagery for immediate inspection as in the MARS type. A laser optical correlation is used to generate image film strips from the holographic film record. The correlator can correct for scale and geometry by referring to recorded flight data and control points. In the GEMS system the imagery is acquired at a scale of 1:400 000, scale changes are effected by enlarging or reducing this source film to the required scale when mosaics can then be constructed using the imagery of the required scale.

Figures 7.7 to 7.9 show the areas of enlargement from the original strip

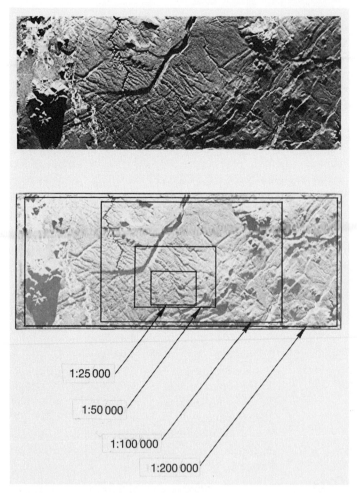

Fig. 7.7   Example of scale variation capability, forested area (courtesy of Goodyear Aeroservices). Scale 1:400 000.

Fig. 7.8   Enlarged section of Fig. 7.7 to 1:100 000 scale.

Fig. 7.9   Further enlargement of section as indicated on Fig. 7.7 to 1:25 000 scale.

and examples of enlarged imagery, the amount of detail on the maximum enlargement is a credit to the fidelity of processing the original film.

It may be appropriate at this point to interject a note of warning in the use of film techniques. It is a fundamental factor in film processing that each successive processing that is undertaken results in a loss of dynamic range and of detail. Every effort is made by the skilled photo-technician to minimize these effects but the factor cannot be eliminated. A film recording system could involve processing signal film, enlarging, mosaicing, copying and reproduction, each step losing some of the original data for the interpreter. Furthermore, in order to produce a balanced mosaic the photographic studio may well introduce photographic cosmetic techniques to achieve a harmonious balance, such techniques will affect the basic interpretability of the data. The techniques include spot reducing the intensity of one image in order to match the lower intensity of an adjacent print, under or over developing and the use of an air brush to merge edges (Figs 7.10 and 7.11). By comparison copying of digital imagery on successive tapes should not degrade the data, enlarging and reducing from digital data may mean that all the data available are used in the enlargement and that sampled data are used in the reductions achieving a compatible data balance at each stage (see Section 5.2).

The Aeroservice processing recognizes the effects of successive photographic processing and accordingly returns to the original signal film for each processing change or enhancement. This factor is of potential importance, a survey flown and optimized for a geological survey may be of subsequent interest to a vegetation study. The enhancement may not however optimize on the data content for vegetation having been enhanced for geological interpretation. Returning to the signal film it is possible to re-work the image in order to improve the expression of the vegetation.

An interesting technique has been introduced into the processing of GEMS radar data which uses a colour film instead of black and white film. By relating the signal return to colour as well as to dynamic range a coloured image is produced which can be likened to a false colour composite or a form of density slicing. It is claimed that such a display illustrates more of the radar data and improves interpretability, certainly the result is perhaps more pleasing to the eye and the use of colour relates the image more to human experience.

The GEMS system has been used with considerable success on numerous studies within the USA and many surveys in other areas throughout the world. Perhaps the most notable of the GEMS surveys was the RADAM project in Brazil, with this survey and subsequent surveys most of Brazil has been covered by radar. Other surveys have been conducted in Indonesia, the Philippines, Colombia, Bolivia and Australia. Some of these projects are described in the following chapter on application.

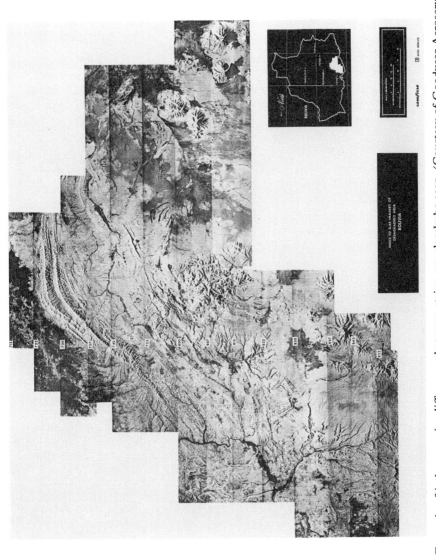

Fig. 7.10 Example of index mosaic: differences between strips can clearly be seen. (Courtesy of Goodyear Aeroservices.)

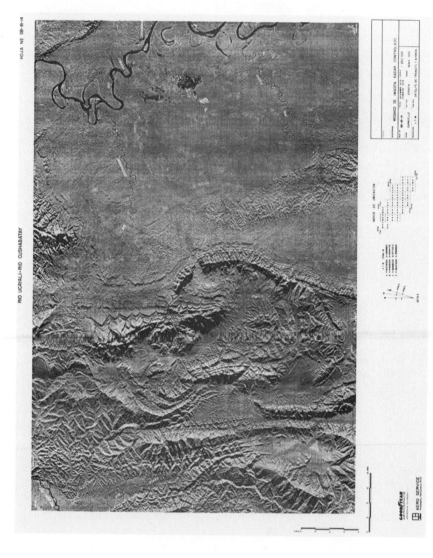

Fig. 7.11 Example of final radar map sheet. The strip differences are no longer apparent. (Courtesy of Goodyear Aeroservice.)

7.3.4   SAR-580

One of the most advanced systems built in recent years has been that built by the Earth Resources Institute of Michigan (ERIM), not strictly as an operational radar, it was designed as an experimental SAR capable of digital recording and imaging in more than one frequency. Table 7.4 shows radar operations carried out by ERIM from 1975–80, not necessarily with the same SAR instrument however.

The system was subsequently operated by Intertech Remote Sensing Ltd of Ottawa Canada, largely on behalf of the Canadian Centre for Remote Sensing; it has now been returned to ERIM and is being operated by the US Navy for oceanographic research.

The aircraft originally used as a platform was a Corvair 580 owned by CCRS and the whole system was used during the Canadian Surveillance Satellite Program (Sursat) the SAR being used to simulate the performance of a satellite borne SAR.

The radar can be adapted to operate in X-band and either C- or L-band and in two polarizations in each of the two selected frequencies. In order to change from C to L or vice versa an antenna change is required as well as modifications to the transmitter. Figures 7.12 and 7.13 show the instrumentation set up in the aircraft. The amount of electronics involved are at once apparent, as this is to a large extent an experimental radar no attempt has been made to package the instrumentation which could result in a reduction of the equipment required.

The radar antennae were mounted in a radome aft of the wings, the X-band consisting of a horn with horizontal and vertical polarization feeds, the L-band consisting of a phased array with two polarizations and the C-band of two separate horns also with two polarization feeds. The antennae are hinged such that a variety of depression angles can be set.

The system is capable of recording on optical film records, recording all four channels of signal data onto film (i.e. 2 bands × 2 polarizations), it has a high density digital tape (HDDT) recorder which will record any two channels as selected, there is a RTD laser recorder for recording digitally processed X-band data and finally there is a strip chart recorder for recording navigation and SAR performance parameters.

The system was used for a number of campaigns, the majority have been in Canada (Figure 7.14) with a subsequent campaign throughout Europe for the European Space Agency and in 1983 a comparable campaign in Japan. In all cases a variety of imagery was obtained for researchers in a wide spectrum of disciplines. The C-band was extremely experimental and not normally used on the aircraft which was also operated on a regular basis for the Canadian Centre for Remote Sensing in an operational mode for ice surveillance.

Table 7.4 ERIM synthetic-aperture radar data gathering missions 1975–80

Dates	Location	Purpose
April 5, 1975	Monroe County, Michigan	Flood assessment
October 7–12, 1973	Brevard County, Florida	Floating vegetation and marsh land detection
October 7–12, 1973	Brevard County, Florida	Urban land use – Titusville
October 7–12, 1973	Brevard County, Florida	Rural land use – south of Cocoa
October 17, 1973	Southeastern Michigan	Agriculture
March 13, 1974	Whitefish Bay, Michigan	Ice
April 1–5, 1974	Phoenix, Arizona Area	Soil moisture, agriculture
April 18, 1974	Oconee County, Georgia	Hydrology, agriculture
April 18, 1974	Southeastern Kentucky	Strip mines
October 30, 1975	Western Michigan	Search and rescue
November 10–11, 1975	Oklahoma City, Oklahoma	Hydrosat
December 3–15, 1975	Marineland, Florida	Seasat – Wave imagery
March 18, 1976	Georges Bank, Massachusetts	Fishing vessel, surveillance
February–March, 1977	Newfoundland, Canada	Sea – ice transit
March 1977	Fort Hood, Texas	Imagery
October, 1977	Lake Michigan	Wave studies
October, 1977	West Virginia	Geology study
July, 1978	Michigan	Demonstration
July, 1978	Ottawa	Agriculture studies
July, 1978	Vancouver	Wave studies
August, 1978	Ottawa	Agriculture studies
August, 1978	New Brunswick	Forestry, agriculture
August, 1978	Gulf of St. Lawrence	Wave studies
August, 1978	Nova Scotia	Agriculture studies
September, 1978	Baffin Island, Thule	Oil – ice studies
September, 1978	Vancouver	Wave studies
October, 1978	Ottawa	Agriculture studies
November, 1978	Atlantic ocean	Oil spill
November, 1978	Ottawa	Agriculture studies
March, 1979	Beaufort sea	Sea ice studies
April, 1979	Greenland and Labrador	Ice and icebergs
May, 1979	Winnipeg	Flood assessment
September 16–18, 1979	New London, Conn. Aberdeen, MD., Andrews AFB	Imagery
October 1–2, 1979	Michigan	Demonstration
October 4–6	Thule, Bylot Islands	Sea ice studies
October 17, 1979	Fort Hood, Texas	Imagery
October 18–23, 1979	Nevada, Arizona	Geology studies
February, 1980	Michigan	Demonstration

Fig. 7.12   The SAR-580 system installed in the Convair-580 aircraft.

In order to demonstrate the complexity of a SAR configuration Fig. 7.15a and 7.15b have been included. These show, respectively, the system block diagram with three image type outputs and the receiver/transmitter block diagram which shows how the X-band is constant and the L- and C-bands are interchangeable.

Figure 7.16 reproduced from the SAR-580 system specification illustrates the complex geometry for motion and navigation geometry that have to be compensated for. The aircraft used an inertial navigation system mounted

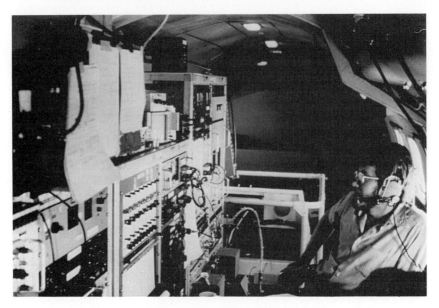

Fig. 7.13   Another view of the SAR-580 installation.

directly over the antenna which supplied the radar with data on ground speed, desired track combined with wander angle, track angle error with drift angle and the three acceleration components for three perpendicular axes. The SAR-580 recorded digitally onto high density digital tapes (HDDTs) using a system developed by ERIM. It also incorporated two other recorders, a precision optical recorder and a real-time processor (RTP) developed by MacDonald Dettwiler and Associates Ltd.

Descriptions of the European SAR-580 campaign are given elsewhere in this book, suffice to say that with all its faults and it had quite a number, the SAR-580 provided considerable essential research data, data on the comparison of frequencies and polarizations for example that had not been available previously, it also introduced a large number of earth scientists and others to the potential of radar.

7.3.5   IRIS and STAR-1

Experience with the SAR-580 can be said to have led to the development of the IRIS, a SAR system built by the MacDonald Dettwiler Technologies of Canada. The STAR-1 is essentially an IRIS in X-band built specifically for Intertech and as such is operated by them, the general configuration can therefore be said to apply to both systems.

Fig. 7.14 ERIM Intera SAR-580 multiband system. X-band, HH like polarized imagery of Ontario, Canada. (Courtesy of Intera.)

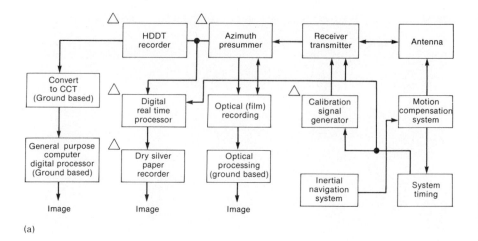

(a)

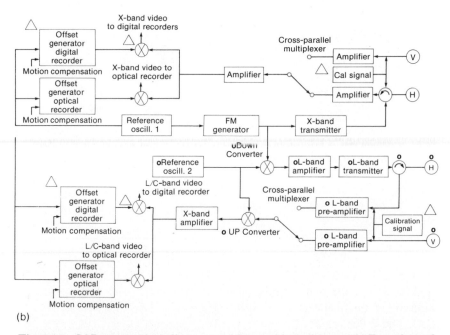

(b)

Fig. 7.15 SAR system block diagrams. (a) System block diagram. (b) Four channel receiver–transmitter block diagram. Δ developed during Sursat O to be replaced by C-band components in the C-band mode.

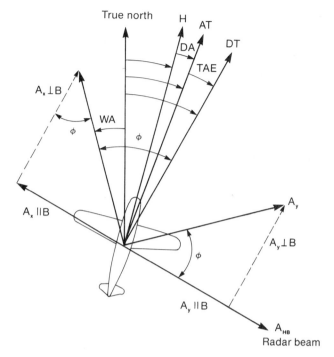

Fig. 7.16   SAR-580 navigation and motion compensation geometry.

TAE: Track angle error $= DT - AT$.

DT: Desired track.

AT: Actual track.

DA: Drift angle $= AT - H$.

H: True heading.

WA: Platform heading defined in CCW direction.

A_x, A_y: Horizontal acceleration components.

ϕ: $WA + DT$.

One of the main advantages of the IRIS is that it has been designed and built as an operational unit, suitably packaged to fit into a wide variety of aircraft. In this respect it differs greatly from the complex package of the SAR-580 which required a substantial aircraft to house the equipment. The IRIS is a 500 kg payload requiring a power supply of 400 Hz, 30 115 V, 5 kW. In its basic configuration IRIS is an X-band SAR with a slant range resolution of 6 m in high resolution and 18 m in wide swath, azimuth resolutions of 6 m and 10 m, respectively, for high resolution and wide swath with a swath width that can be either 20 km or 60 km depending upon the mode chosen.

The IRIS can also be configured to operate in the C-band mode. It has on-board digital processing and image production facilities, optical real time products using either a dry silver paper or film transparency. The system can operate either in a wide swath mode or high resolution mode and the switch from one mode to the other can be made in flight.

Although on-board digital processing can be supplied it is also possible to record on HDDTs on board and carry out precision processing subsequently at a ground station in order to obtain maximum precision data. In certain operations it is essential to be able to relay real time data to a ground station, for example an ice breaking ship looking for leads in the ice pack or an anti-pollution vessel seeking an oil spill. The IRIS can be transmitted to a ground station if required.

Figure 7.17 shows the STAR system fitted into a smaller aircraft, compare this diagram with the earlier configuration of the SAR-580 and the improvement is readily apparent.

The IRIS as the STAR-1 (Table 7.5) is being used operationally in Canada and the vastly improved image quality is finding considerable favour. A C-band system is to be operated by the CCRS as part of their development work for their forthcoming satellite radar (see Chapter 12), Fig. 7.18 is an example of STAR-1 data.

Table 7.5   STAR-1 performance specifications

Parameter	Design specification	
Operating altitude	10 km	
Wavelength	X band	
Polarization	HH	
Viewing direction	Left or right	
MDS	-30 dB at maximum range	
System-related image modulation at downlink display	less than 3 dB in range and azimuth	

	Design specification	
	High resolution	Wide swath
Ground swath width	25 km	50 km
Maximum range	40 km	60 km
K =	5	10
Range resolution	6 m	12 m
Azimuth resolution	6 m	6 m
Azimuth looks	7	7
Range looks	1	1

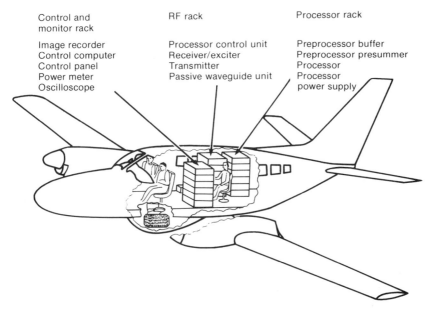

Control and monitor rack

Image recorder
Control computer
Control panel
Power meter
Oscilloscope

RF rack

Processor control unit
Receiver/exciter
Transmitter
Passive waveguide unit

Processor rack

Preprocessor buffer
Preprocessor presummer
Processor
Processor
power supply

Fig. 7.17   STAR-1 aircraft configuration (after Intera Technologies).

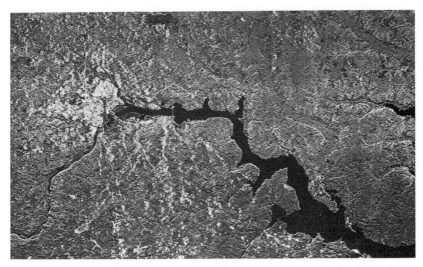

Fig. 7.18   Example of STAR-1 imagery. (Courtesy of Intera Technologies.) SAR image of Washington DC. Frequency X-band; scale 1:250 000; resolution 6 × 12 m; swath width approximately 50 km.

7.4 ADVANCED EXPERIMENTAL SYSTEMS

A number of systems have been built for research applications. The Jet Propulsion Laboratory of Pasadena, California, for example, built a 25 m resolution L-band SAR which was flown in a NASA aircraft and used on a number of resources surveys in arctic regions. In France, Thomson CSF have built systems, one of these is a SAR in X-band with 30 m resolution that has also been used successfully on a number of missions. In England systems have been built and installed in military experimental aircraft and in this respect it has to be recognized that more advanced systems do exist but these are all used in military applications and details are not available.

In Holland however the Dutch radar research group ROVE, operate their own X-band SAR for research mainly into crop and vegetation classification. This has been used for annual campaigns in 1982, '83 and '84, with multi-temporal imaging throughout the growing season. The instrument can vary its angle of depression and has a range resolution of 7.5 m. At the same time an airborne scatterometer is used to extend the range of measurements.

8

Geological interpretation

8.1 INTRODUCTION

This chapter describes a number of projects that have been undertaken with geological interpretation as the major consideration. Other resources studies are dealt with in a similar manner in subsequent chapters. It is inevitable that some surveys are multi-disciplinary and where this is the case the description may appear in each chapter, as appropriate to the discipline.

Geologists have always been one of the principal users of airborne data, some of the earliest thematic interpretations of air photographs were undertaken for geological surveys. This is understandable when it is considered that the structures of the earth's crust may extend over many miles and not be readily identifiable in their entirety from the ground. Air photographs and mosaics provided a synoptic view which enabled these structures to be observed and mapped for the first time. The advent of satellite imagery was of great importance to the geologists; now an area of approximately 100 miles × 100 miles (160 km × 160 km) could be observed on one single picture. This was of significance because mosaics constructed from smaller air photographs, although controlled to a map base, may show slight variations in orientation between images which, although not apparent in the local mosaic context, could well mask a major significant linear trend. This proved the case in a number of instances, areas previously mapped from air photographs have contained unexplained structural anomalies for which seemingly logical explanations were derived. The Landsat image of the same area revealed a structure that had not been obvious and which explained the previous anomalies.

Geologists have also used other airborne systems with the use of airborne geophysics in their search for another view of the earth. Airborne radars offered a number of potential advantages, the all weather capability was an obvious attraction in observing areas for which imagery had not previously been available. The side looking character of radar and its strong shadows meant that structures were well displayed, including minor structures formed from poor relief, this in turn led to the realization that linears could be well displayed.

Some of the systems, such as the Shuttle series SIR-A and SIR-B, had the added advantage, especially in forested areas, of appearing to remove or to even out the effects of surface vegetation. Finally under certain circumstances a degree of penetration was possible, whether this was of vegetation cover or soil layers it meant that features not otherwise visible could be observed.

8.2   SYSTEM PARAMETERS

8.2.1   Frequency

In general, frequency does not play such a significant role in geology as it does in vegetation surveys. The major commercial airborne systems have been Ka- or X-band, the majority being the latter and most of the surveys, especially those in the USA, have been for geological studies. Subsequently the shuttle series have used L-band and these flights have had a large geological objective. The longer wavelengths are of interest because of their greater penetration potential and this has proved of significance in studies over desert areas where the longer wavelength has helped to reveal subsurface drainage patterns.

8.2.2   Polarization

Where dual polarized imagery has been available it has not made a significant difference to geological presentation and most researchers advocate HH only.

8.2.3   Incidence angle

The ability of radar to illuminate topographical features to provide an exceptional presentation of relief is of greater interest to the geologist. The degree to which minor structures can be highlighted on an image is a factor of the incidence angle, however in areas of severe and high relief the incidence angle suitable for subtle features will produce large unacceptable radar shadows often obscuring other valuable data. Furthermore, although an increase in depression angle will decrease the shadow area it will at the same time increase the possibility of layover in severe terrain.

Figure 8.1 relates radar response with incidence angle for a vegetated surface, where it can be seen that for incidence angles smaller than 30° the radar response increases with 0.5 dB per 1° decrease of angle.

At the same time the incidence angle varies across the swath so that the far range may show areas of low relief which if observed in the near range may not be recorded.

The SIR-B shuttle SAR mission was designed to obtain imagery at a variety of depression angles in an attempt to identify the optimum angle or angles for geological studies. Ideally for geological study the SAR would be acquired with different angles but in the case of airborne data, this may not always be practical. It is therefore essential to study an area carefully before survey in order to choose the best depression angle or possibly to segment an area into zones using a different angle for each zone.

8.2.4 Look direction

Since the geological survey will almost certainly be concerned with structures and the identification of linears, look direction can be of paramount importance. The look direction should be designed with the grain of the country in mind and this has already been noted in Chapter 6 on field work and planning.

The choice of look direction cannot be selected with space imagery, the direction will depend upon the orbit pattern of the platform and this has to be accepted as being potentially non-ideal for many areas. On the other

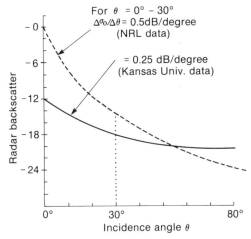

Fig. 8.1 Effects of varying incidence angle on radar backscatter (after Ford.)

hand, the imagery can be acquired in both ascending and descending mode when the inclined orbit will result in different look directions from the two orbits. Attention is called to the linear bias of radar (Fig. 2.20), the correct look direction is essential to minimize this effect which is further reduced with more than one look direction. The dual look flight planning or data acquisition as used by MARS is desirable in order to counter the effects of shadow, this does not however necessarily remove the linear bias when a flight at 90° would be preferable.

8.2.5   Resolution

Compared with agriculture surveys or urban mapping, geology is not so critical on resolution. Having said that, it has to be realized that there are many different branches of geology and what may be suitable for one may not be optimum for another. In the broader regional geological surveys a fine resolution is not only not necessary but may even be undesirable. There is an argument which says that one of the benefits of SAR is the smoothing effect on the surface features (vegetation etc.) enabling the structures to predominate.

On the other hand, more detailed local surveys such as that carried out on an open cast mine in Spain under the European SAR-580 campaign (Koopmans, Pacheco, Woldi and Payas, 1983 and 1985) would require the higher resolution in order to resolve smaller structural details.

8.2.6   Multi-temporal

Multi-temporal imaging is, again, not so critical for geologists as it is for other disciplines and there is a strong opinion among many geologists that if the earth is imaged in detail just once this will satisfy most of their requirements. For regional mapping this is probably true, but geology is not static and many of the structures are still actively forming and other geological events occurring. This once more depends upon the branch of geology concerned, for example the engineering geologist may well require regular imagery over an area in order to monitor the effects of land slides; a study in Calabria, Southern Italy using SAR-580 data (not reported in the literature) has been concerned with this very problem in a region noted for its major landslides and the resulting problems.

Major geological events such as volcanic activity would also benefit from regular monitoring but these would be difficult to plan except in the most active zones, mainly because volcanic events at any one site may occur only after a long non-active period, e.g., Mount St Helens (Fig. 8.2).

In general therefore multi-temporal data are not of major geological interest although they may be a requirement in local areas.

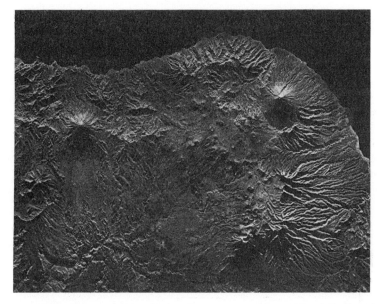

Fig. 8.2   The radar expression of volcanic structures. Flores Island, Indian Ocean. (Courtesy of Aeroservice.)

8.3   TARGET PARAMETERS

In other disciplines, as for example agriculture, forestry, snow and ice and oceanography, there has been a considerable amount of research into the interaction of targets with microwaves. This has taken the form of theoretical modelling, the use of scatterometers in regulated measuring and the study of actual imagery.

There is much less comparable research in geological subjects except insomuch as elements of the other disciplines affect geology, this may for example be the relationship or dependence of vegetation on the underlying surface materials, or the observation of subsurface features from microwave images of sea areas. These other relationships should not be minimized, the ability eventually to discriminate vegetation types has been shown to relate to underlying geology, indeed geological mapping in the large forested areas has to rely on such indicators to determine the possible lithology.

In areas where there is no vegetation cover, whether soil or bare rock, then the vital parameter is the surface roughness and this is one area which has been the subject of perhaps the major research from Rouse *et al.* (1966) to MacDonald and Waite (1973) through to more recent studies. In an area of exposed rock surfaces, the texture of each lithological unit may be the only available discriminator since radar cannot record possible colour differences. The effect of surface roughness will vary with the frequency used, the angle of incidence and the angle the surface presents to the viewing platform. Koopmans (1983) has prepared a graph, Fig. 8.3, to show the effect of incidence angle against vertical relief for a number of frequencies.

In 1978 Sabins produced a prediction of the theoretical signatures against observed signatures for rock units in the Cottonball Basin USA. These are shown in Table 8.1.

The soil moisture content, or dielectric constant, is recognized as being a major influence on the radar reponse. Different rock units have equally varying capacities to hold and retain water, some rocks are extremely porous and may show a high water retention. This in turn will affect the backscatter return and different rock units may be determined by their dielectric constant. This ability will be in addition to the surface roughness value for a

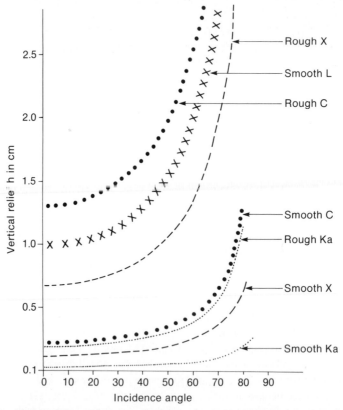

Fig. 8.3 Height differences of surface irregularities indicating smooth or rough surface limits against radar incidence angles for different wavelength bands. (From Koopmans 1983.)

Ka-band: $\lambda = 0.8$ cm; X-band: $\lambda = 3.0$ cm; C-band: $\lambda = 5.7$ cm; L-band: $\lambda = 25$ cm.

Table 8.1 Observed and predicted signatures

Radar rock unit	In average (measured)	Observed signature	Predicted signature
Flood plain deposits	0.2 cm	Dark	Dark
Desert pavement	1.0 cm	Intermediate	Intermediate
Carbonated-cemented sand	6.0 cm	Intermediate	Bright
Smooth rock salt	6.0 cm	Intermediate	Bright
Coarse gravel	12.0 cm	Bright	Bright
Rough, eroded rock salt	730.0 cm	Bright	Bright

rock, the presence of high moisture content has been found to explain the two differences in Sabins' table.

The moisture content of surface materials will also influence the amount of penetration achieved by a microwave pulse, the penetration of desert sands to show sub-surface drainage in the North African desert noted on the SIR-A Study (Elachi, 1982) has been largely attributed to the very dry overlying material permitting penetration.

The advent of satellite radars will result in two overpasses, one will be in daylight and the other at night, this may well influence radar signatures, the night time image may record a high dielectric constant whilst the daytime image may be lower due to evaporation resulting in a different signature for the same unit.

8.4 STEREOSCOPY AND SLOPES

The expression and interpretation of relief on the single image relies entirely upon the shadow effect; the variations in incidence angle across the swath and other factors mean that whilst structures can be detected and their influence inferred, no true relative heights can be derived nor dip slopes estimated. Ideally therefore the geological interpreter would like to have the ability to study an area stereoscopically. The problems of stereoscopy in radar have already been discussed and the ability to make true measurements is not yet viable.

Despite the fundamental geometric problems it is possible, given sufficient overlap, to study radar images stereoscopically and where this is possible the interpretation from such a view is improved and the level of confidence increased (Koopmans, 1974). In theory the north look and south look images of the same areas should provide for a stereoscopic view since the equivalent of two eye bases is used. In fact in areas of relief the opposing shadows tend to eliminate large areas of detail with only the common mountain ridge in view, so that the mental reconciliation of the two views is virtually impossible.

In studying an image for geological interpretation it is important to obtain an assessment of the slope of the beds, the dip slopes. A number of factors make an accurate assessment almost impossible from radar.

The effects of layover and foreshortening of sloped surfaces on a radar image were discussed in Chapter 2. These elements become of significance in geological interpretation. In the extreme case, that of excessive layover occurring, it is clearly impossible to make any assessment of the slope involved. In other less severe cases a slope dipping towards the line of flight will appear shorter than a comparable slope at the same angle dipping away from the line of flight.

A typical set of inclined beds with a dipping back slope and a steep forward edge will be best imaged by flights planned to be parallel to the line of the beds, but consideration must also be given as to whether the look is to the front slope or to the back dip slope. Furthermore the foreshortening effect will vary with the position across the swath.

Scarp slopes facing the platform, particularly those with broken surfaces, will result in a high radar response on the image which will possibly be disproportionate to their importance. The back slope will not give such a strong return. A structure such as a syncline with a parallel set of structures facing the opposite direction will still not give such a bright return on the back slope (now facing the platform) whilst the steeper scarp may appear as a thin shadow line. An interpreter could therefore assume that the slope with the scarp facing the platform was steeper and more significant than the opposite slopes with the scarp facing away. Figure 8.4 shows how structures can be wrongly interpreted from the apparent relief data.

8.5 DRAINAGE

One of the most significant interpretation features on a radar image is the drainage, it is usually clearly defined and relatively easy to map and it has a significance in helping interpretation for nearly every discipline, none more so than geology (Fig. 8.5). The drainage of an area is controlled by degree of slope, permeability of underlying materials and by the ease with which the

Fig. 8.4   The potential misinterpretation of geological structures based upon the radar representations of relief.

Fig. 8.5 Drainage patterns clearly represented on SAR. The Altiplano, Bolivia (courtesy of Aeroservice).

water can cut itself channels. These channels may in turn be formed from existing structural features such as faults. Changes in surface conditions will be reflected by changes in the drainage pattern. The drainage also causes changes in the geomorphology by incising valleys, by depositing eroding materials to form alluvial plains and then meandering through them in ever changing courses and by forming the large delta features.

It is therefore a first requisite on many images to map or delineate the major drainage features if they are not immediately obvious. Boundaries can then be drawn around different drainage types and after modification by any other factors (vegetation changes, field data etc.) these usually serve as a basis for geological interpretation. Figure 8.6 is an area in which the units have been mapped solely on the basis of the drainage patterns.

Reference has been made to the smoothing effect of radar on vegetation cover in high forest regions. This was particularly noted during the radar surveys of Colombia and Brazil. There is a further effect however: the presence of vegetation appears to enhance rather than obscure the representation of drainage systems. Drainage which causes a cut in the tree canopy is enhanced by that break, the trees facing the imagery platform seen as edges as well as canopy will give a high return, whilst the other edge facing away from the sensor will result in a shadow. In wide floodplains the

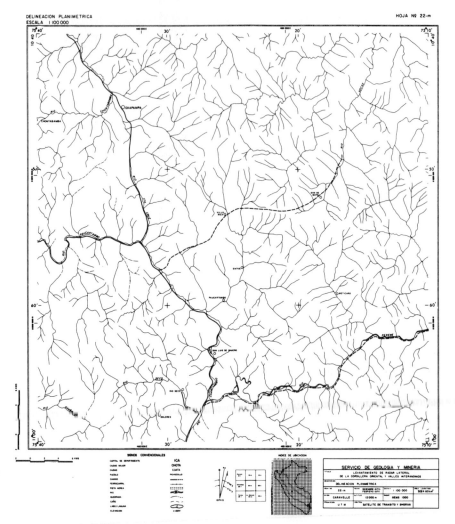

Fig. 8.6   Example of planimetric drainage overlay from SAR imagery (courtesy of Aeroservice.)

different meanders, ox-bows and lakes can be clearly observed and where present the change from flood plain to sloping drainage surface is usually well defined.

8.6 LINEAMENTS

By now it must be obvious that the recording of even minor structural or linear features is one of the major advantages of SAR over other image systems. This is of particular importance in geology where the interpretation, mapping and analysis of linears plays a major role in geological mapping.

The regional distribution of faults and linear patterns in western Irian Jaya, Indonesia, was first seen on airborne radar acquired in 1974 with an X-band radar. Analysis of the regional structural pattern suggested that plate tectonism was responsible for the fragmentation that has separated the oil-producing Salawati Island from the mainland (Froidevaux, 1978 and 1980).

When interpreting radar linears, the bias of the system has to be borne in mind, wherever possible mapping from two look directions is preferable. The nature of radar is such that all the reasons for linears may not be obvious, thus artificial features such as roads and railways, under certain circumstances, may be misinterpreted as linears. This may be both a weakness and a strength, the latter because all linears are usually given equal weighting and no prequalification made. In some cases roads etc. following the easiest natural route do in fact follow natural geological features.

The pattern and analysis of linears is a technique in its own right and it is not the remit of this book to pursue this particular subject. It is sufficient to say that the intersection of major linears can indicate possible mineral accumulations and Gelnett and Gardner (1979) have shown that a study of the linears in a basement rock area in North Nigeria, by locating the intersection of major feature patterns, was highly successful in guiding drill teams in their search for valuable ground water.

Linears are most apparent when they are topographic features such as valleys or scarp edges and it is this form of linear that is particularly subject to the linear bias (note Landsat has a similar bias). The main result can be that a linear can assume greater importance than it deserves because of the strength of its representation, while a major linear at right angles to the line of flight can have no expression and be overlooked. Changes in material along a fault zone can result in a change in dielectric constant as water is retained in, say, breccia material, this could give rise to a change in return identifiable as a linear feature on an image. Vegetation too can reflect underlying linears and these may not be so affected by directional bias.

8.7 DIGITAL ANALYSIS

The majority of image analyses for geological studies are carried out using conventional image analysis techniques. Where digital data have been available they have largely been used to produce an improved photographic product for analysis. This is natural when most of the interpretation has been of lineaments, structures and drainage analysis – interpretation methods best carried out by an interpreter directly on the image.

In the study in Spain, carried out on SAR-580 (Koopmans et al., 1985) digital analysis was attempted but the image data were so poor that it failed to improve interpretation. This is an instance when tonal variations were being studied in order to differentiate units.

The problems of geometrical restitution remain and these are serious where a geologist wishes to relate data to height information, particularly in hilly areas.

Digital techniques have been used to combine a radar image to a Landsat image, this has proved successful because the better relief representation given by the radar improves the interpretability of the total scene.

The effect of surface roughness for each frequency and its relationship to lithological mapping, where beds are suitably exposed, does hold the possibility of using radar in combination to produce a colour composite that could discriminate between units. As far as can be ascertained this has not as yet been attempted.

8.8 VENEZUELA

The radar imagery of Venezuela has been described as the first major SAR survey in South America, it certainly was the largest single survey of its time involving over 210 000 km^2 of territory.

The Venezuelan Government wished to survey the State of Amazonas, a mountainous area covered in rain forest and invariably under cloud cover making it impossible to survey using optical methods. In this instance it is interesting to note that throughout the entire survey, the aircrew were unable to see the survey area at any time from the aircraft. Under such circumstances air survey visual techniques are of no use whatsoever.

The subsequent analysis of the imagery was undertaken for a variety of disciplines but in the geological study maps were constructed to show the structural geology, the stratigraphy and to indicate the potential economic geology of the area, from which recommendation for the possible development of mineral resources could be made.

Part of the study was concerned with the geomorphology of the region and in this context the river systems were identified and mapped to provide a picture of the drainage network. Out of this the interpreters were able to identify a number of potential hydro-electric power dam sites, six of which were sufficiently encouraging to warrant more detailed investigations. The study of the drainage pattern also located a river that was previously unknown. At the end of the study an area of previously unknown potential had become an area with a potential for hydro-electric power development and, perhaps of greater importance, an area with potential for exploiting economic minerals and which could be developed for forestry and farming.

The study was eventually extended to the entire country, an area of 900 000 km^2.

8.9 PRO-RADAM

The *Projecto Rad*argrammetrico del *Am*azonas (Pro-radam) carried out over Colombia is described under Forestry in Chapter 9. It is mentioned here as it was a major project that also included geological interpretation (Fig. 8.7).

8.10   THE RADAM PROJECT

Perhaps the most famous of all the radar surveys is the RADAM (*Rad*ar na *Am*azonia) project carried out in Brazil for the Ministry of Mines and Energy in 1970 and covering the Amazonian area in the northern part of Brazil (Figure 8.8). The survey was conducted using the Goodyear/ Aeroservice SAR and under the technical supervision of the Earthsat Corp. Although the contract was commissioned by the Ministry of Mines and Energy, the studies ranged over all resources: geology, geomorphology, soils, present land use, vegetation and land use potential. The output was a number of detailed reports and a series of maps for each of the survey elements. The objective was described (Bittencourt Netto, 1979) as to

Fig. 8.7 Barranquilla imagery segment of Colombia, South America. The mosaic depicts features of the Magdalena River as it flows northward past the city of Barranquilla (left). Roadways and agricultural patterns span the areas between smaller towns. The imagery for this mosaic was recorded by side looking airborne radar from 40 000 ft (12 km) altitude through total cloud cover. The area represented is approximately 28 × 37 statute miles (45 × 60 km).

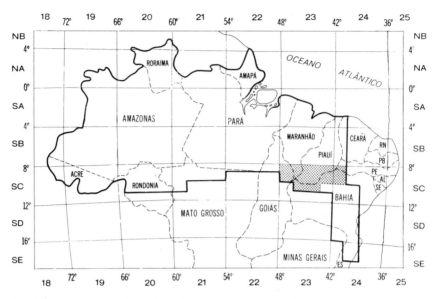

Fig. 8.8   Area covered by the RADAM survey in Brazil.

'update and organise data concerning Brazilian Amazonia, with a view to providing national guidance for its settlement and exploration'.

The nature of the country surveyed (large tracts of dense tropical forest) meant that survey by conventional field work would be both time consuming and hazardous and to all practical purposes, impossible. At the same time the presence of high forest is indicative of high rainfall conditions which in turn indicate a preponderance of cloud cover. Thus Landsat would not offer an immediate solution as a means of obtaining images and radar was the obvious solution. Landsat has since been used, indeed Brazil was one of the first countries outside North America, to install its own Landsat receiving station. With the base line survey provided by radar it is possible to upgrade and extend the interpretation by using Landsat whenever a cloud free or partially cloud covered image is available. This application points to an important application of radar, it may not present all the data required for interpretation but it can provide a valuable base line survey which can be revised as other data become available, planning can be conducted without delay using the base data and not hindered by data collection constraints.

The field work in such remote and impassable areas provided particular problems. Eventually recourse was made to specialized teams trained in jungle survival, who went into the forest by foot, by river and even by being lowered on ropes from helicopters, in order to clear forested areas and set up base camps for the teams of scientists involved. The usual method of survey in such circumstances is to cut radial trace lines out from the camp site and survey along these traces. In such circumstances the field of view is strictly limited and the teams could not obtain a view far beyond the trace lines.

The geological survey was able to identify the Mesozoics and Basalts in the region, to improve the existing knowledge of the stratigraphy and structures and in some cases to correct the existing data. In particular the survey was able to indicate the location of major mineral occurrences with suggestions of promising areas for future prospecting and survey including identifying structures which could be of potential interest in oil exploration. Figure 8.9 is an example of the type of geological interpretation overlay produced.

The geomorphological studies were complementary to the geological studies as would be expected, the geomorphological and geological studies in turn are of significance to the soil survey and are used to determine areas

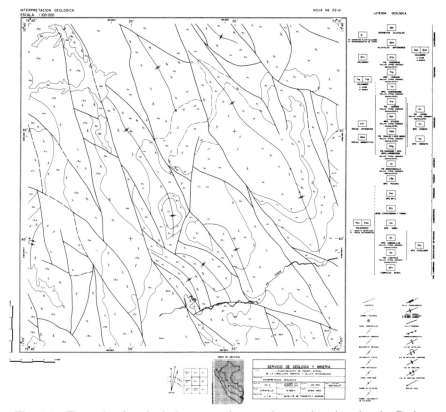

Fig. 8.9 Example of geologic interpretation overlay produced under the Radam project. The key is included in order to give an idea of the degree of detail shown.

of land use potential. Thus the study highlights the inter-dependence of disciplines and the need to co-ordinate studies and it is as a result of these studies that Brazil set up its remote sensing centre to co-ordinate resources studies.

The radar image has already been shown to be ideal for displaying the relief of an area despite the heavy vegetation cover. This capability of the radar assisted in defining the geomorphological units of the country. The precise mapping of soils requires a great deal of field work, soils can only be properly classified by studying the soil profile in dug pits and by systematic sampling of the soil at varying depths using an auger. Part of the classification is derived from field observations: texture, colour, associated vegetation, relationship of horizons etc.; and part from chemical analysis of the collected samples. Clearly these data cannot be obtained from radar therefore these soil units are mapped as associations being related to geomorphology and vegetation units. The soil maps in Radam were derived by using these criteria supplemented by existing survey data or additional field work to enable a reliable extrapolation to be made.

The survey area covered large tracts of forest lands but not all the area was so covered, there was also woodland savanna, parkland and existing cultivated lands. In these latter areas field studies were possible, even so in these and the more inaccessible areas use was also made of light aircraft to fly over areas and obtain oblique aerial colour photographs. The climate is clearly a dominating factor in such an area and the interpretation of vegetation units would require access to climatic data as well as the imagery. The vegetation survey, based on tone, texture and contextural analysis of the image, also drew upon the geological and geomorphological features but included reference to climatic data.

The land use potential maps are perhaps not strictly radar maps since they are compiled from an interpretation from all the other interpreted maps, thus similar geomorphological areas may be identified as suitable for agricultural development but the vegetation and soil maps could indicate that they would be suited to two different crops. It is often necessary to refer back to the imagery to clarify a point not perhaps brought out by one of the surveys but which may be vital for crop or land use selection, e.g. proximity of existing infrastructure roads, villages etc. In assessing the potential land use, the effects of development were taken into consideration, for example potential erosion hazards or flood conditions exacerbated by changing the river regime.

The radar data and subsequent Landsat supplementary data have provided Brazil with invaluable information on its natural resources which have enabled the country to plan the development of its natural resources in an efficient manner. It is interesting to note that such was the success of the study that, despite the installation of a Landsat station, the country has since commissioned a radar survey of the remainder of the country for similar study purposes.

8.11 NICARAGUA

In 1971 a radar survey was undertaken of the whole of Nicaragua and this was the first time a whole country had been so imaged. The survey was conducted using the Westinghouse K-band radar and is an excellent example of the potential economics arising out of the use of radar. Nicaragua is a country of 119 000 km^2 containing a varied landscape of mountains, forests and swamps and as would be expected in such a vegetated country, has a high rainfall and therefore regular cloud cover. The cover for the country was obtained in a total of 10 days or 78.6 flying hours, at a scale of 1:250 000; Fig. 8.10 shows the flight lines for the survey, indicating the magnitude of the flying task in difficult conditions.

In order to achieve such a high flying rate, the survey had to be a twenty four hour operation with processing and assessment of the film being carried out throughout the night in order to plan the next day's flying without incurring any delays.

The radar was mounted in a Douglas DC-6B which flew the survey at an average height of 6100 m (20 000 ft) using a Doppler radar and navigational computer to provide drift and ground speed information.

This type of efficient operation is typical of the way in which radar surveys have to be conducted in order to achieve maximum economics of expensive aircraft, crews and equipment. With radar, however, such stringent flight planning is possible, with the only constraints being severe thunderstorms over the imaging areas or constraints to aircraft operations such as high winds, dust storms or air traffic regulations.

The Ka-band radar, being real aperture, provided a film product which required no further processing and therefore there was no time delay in passing the data to the interpreters for analysis.

The radar data were used to provide an interpretation base for a series of overlays to the mosaics at 1:100 000 depicting geomorphology, geology and land use. The interpretation was for the eastern two thirds of the country and did not cover the whole of the country (Fig. 8.11). The overall objective was to provide basic information for use in assessing the development potential of the region. Although a visit was made to the country by key interpretation personnel, the objective was to collect all relevant data rather than carry out field work. For a number of logistical and political reasons detailed field work was not possible, however some local observations were made.

The survey therefore was very much a reconnaissance survey and in order to improve the results it would have been necessary to include a programme

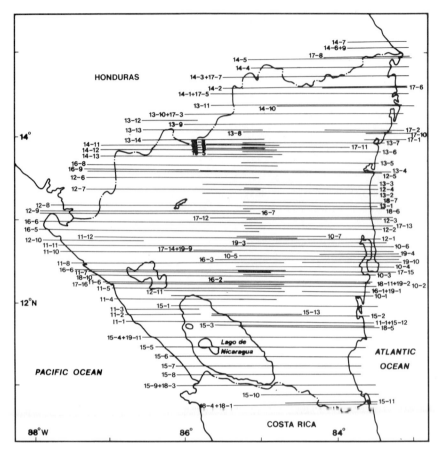

Fig. 8.10 The flight plan for the radar survey of Nicaragua. (Hunting Geology and Geophysics Ltd.)

of systematic field studies, which, given the terrain involved would have been extremely difficult. The survey was able to indicate where there was resources development potential and more detailed surveys would be necessary in order to prove these findings.

The radar image strips were constructed into mosaics and it is appropriate to emphasize that radar images will be subject to many of the problems encountered in conventional photography when mosaicing. The radar oblique view is corrected in the system to give a nominal vertical view but certain geometric distortions will none the less remain; at the same time, in mountainous country, the problem of layover will occur and this will present image matching difficulties when mosaicing. The radar image will also present radiometric imbalance across the swath due to the antenna pattern (see Section 5.2).

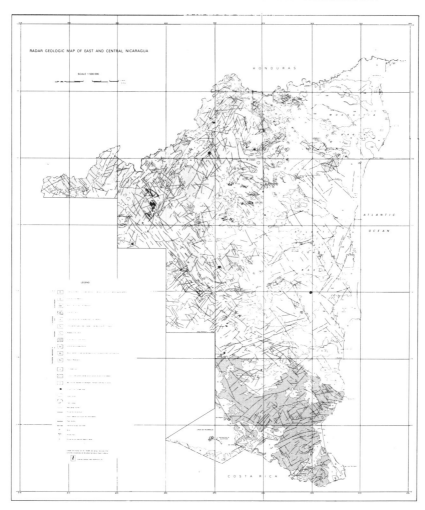

Fig. 8.11 The radar geological maps of Nicaragua. (Hunting Geology and Geophysics Ltd.)

Where there has been excessive correction and cosmetic modification to the imagery in order to construct a mosaic it is preferable not to use the mosaic for interpretation but whenever possible prints made from the original film strips should be used. Where data are optically recorded copies of strips contain the maximum available data. It is possible that in some cases radiometric imbalance may occur across the swath as in the SAR-580 and consideration may have to be given to digitizing the imagery and processing in a computer to minimize the effects but this is an extreme solution and not one generally to be recommended. With the major commercial systems these radiometric differences are not a normal consideration.

These comments only apply to optically recorded data, in the case of digital recording the comments do not apply, especially if the mosaic is digitally constructed.

In the mapping of Nicaragua's morphology and geomorphology, three main morphological units occur, the Atlantic Coastal Plain, the Interior Highlands and the Nicaragua Depression. Within these groups the various drainage basins were delineated and five main categories of terrain recognized: mountain lands, hill lands, flat lands, swamps and wet lands and coastal terrain, with each category having a number of sub-divisions thus:

Mountain lands: massif, broken and rugged terrain and linear ridged mountains;

Hill lands: broken and rugged hill terrain, linear ridged hills, rolling hill land, Cuesta and mesa land;

Flat terrain: undulating plain with low hills, featureless plain, incised low plateau, alluvial tracts;

Swamp and wet land: swamp, silted lagoons;

Coastal terrain: beach lands; and, as a special category, volcanic land forms.

Nicaragua is a country of active volcanoes and the boundary of two major structural units passes through the country, thus the area was of great geological interest (Fig. 8.11). Without adequate field data, and only references to existing reports and maps, much of the interpretation had to rely upon the morphological expression of rock types and structure. In the Nicaraguan geological study about 25 geological units were mapped, together with extensive mapping of the fault and fracture patterns as expressed in linears.

In a country of current volcanic activity the identification of fault patterns is essential to understanding the structure of the country. As a result of the study the existence of a new major fracture zone was identified which could be of significance in predicting future earth movements and is essential knowledge in the planning of future development.

When studying mineralization it was almost impossible to identify old or new mineral workings at a scale of 1:250 000 but applying the knowledge gained of known mineralized areas to the imagery it is possible to identify comparable areas that could be selected for more detailed examination. Thus the survey was able to select priority areas for further mineral investigation thereby concentrating future activities on those areas most likely to yield results.

By studying all the map overlays and applying additional criteria such as accessibility it is possible to make an assessment of the land development potential of a region. In the Nicaraguan survey the report included such an assessment to provide a valuable guide for future development.

8.12 CONCLUSION

In considering geology as with any other disciplinary study it is important to reiterate that the results of the interpretation reflect the other data bases used, these include existing maps, air photos, Landsat, reports and especially field work, however limited this latter may be. Like any other imagery system, radar is a tool for the geologist to use together with all the other methods used for a survey.

The maps showing known airborne radar cover indicate almost all the USA covered at some time by radar. The majority of this has been as a result of contracts requested by the USGS for geological surveys. In 1980 the US Geological Survey began a systematic radar programme and imagery of more than 390 000 km^2 of the main area of the United States and of Alaska has been acquired. The programme has involved Governmental, academic and industrial scientists who participated in more than 50 research and data analysis projects, applying SAR to geology, cartography and hydrogeology. Some of the areas studied already have geological maps and the coloured maps have been used as a colour overlay to the radar mosaic for comparison, the radar mosaic adds a strong structural dimension to the maps and also adds to the data presented. The imagery benefits the mapping of geologic structures especially in the folded and thrust faulted Appalachian mountains. Not all the areas studied have had maps available for comparison and in the Aleutian Islands, Alaska the radar is being used to research into the potential for producing cartographically accurate maps.

In all the USGS work the SAR is finding a role in the mapping of geological structures, in mapping geological hazards such as landslides and in assessing the mineral resources of the region. Radar has, and will continue to have, a vital role in mapping structures' geomorphology and lineaments for geological surveys, it has not been so successful as an aid to mapping lithologies but perhaps the combination of microwave frequencies or polarizations may improve this situation. Some of the future shuttle radar expeditions have geological interests as their primary objective and there can be little doubt of the value of radar to geology.

REFERENCES

Bittencourt Netto, O. (1979) The mapping of natural resources by side-scan radar in Brazil. *Mining Magazine*, 354–359.

Gelnett, R. H. and Gardner, J. V. (1979) *Use of radar for groundwater exploration in Nigeria*. Proceedings of the 13th International Conference on Remote Sensing of the Environment, ERIM, Ann Arbor.

Elachi, C. H. (1982) *The SIR-A sensor and experiment,* IGARSS, 2, FA-6, June 1–4.

Froidevaux, C. M. (1978) Tertiary tectonic history of Salawati area, Iriandaya, Indonesia. *AAPG Bulletin,* **62**, (7), 1127–50.

Froidevaux, C. M. (1980) Radar, and optimum remote sensing toll for detailed plate tectonic analysis and its application to hydrocarbon exploration in *Radar Geology : An Assessment,* JPL publication 80–61.

Koopmans, B. N. (1974) Should stereo SLAR be preferred to single strip imagery for thematic mapping. *ITC Journal,* 424–45.

Koopmans, B. N. (1983) Side looking radar, a tool for geological surveys. *Remote Sensing Reviews,* **1**, 1, Harvard Academic publishers.

Koopmans, B. N., Pacheco, Woldi and Payas (1983) *Sarthi, a side looking radar survey over the Iberian pyrite belt.* The European SAR-580 Investigators' Preliminary Report September 1st, The Joint Research Centre ESA (European Space Agency).

Koopmans, B. N., Pacheco, Woldi and Payas (1985) The European SAR-580 Investigators' Final Report.

MacDonald, H. C. and Waite, W. P. (1973) Imaging radars provide terrain texture and roughness parameters in semi-arid environments. *Modern Geology,* 4, 145–58.

Rouse, J. W., Waite, W. P. and Walters, R. L. (1966) *Use of orbital radars for geoscience investigations.* CRES Technical Report 61-8, University of Kansas.

9

Applications in forestry

9.1 GENERAL

If geology is seen to be the current prime interpretative user of radar imagery, forestry must run it a close second. If one considers that much of the world's forest resources are unmapped and that by definition the large areas of wet tropical forest, i.e. areas with a high rainfall, are under persistent cloud cover, it is easy to understand why the potential of radar is keenly regarded. Even in the northern areas the forests lie in countries under severe cloud cover such as Canada, Scandinavia and the USSR.

9.2 TYPE OF SURVEY

Forest surveys can be divided into two different major groups, those surveys concerned with estimating the extent and general mix of a forest and those concerned with managing known plantations. In the former case the first requisite is to map the extent of forest reserves and then to monitor their preservation or exploitation, usually on a very large scale where monitoring would be difficult.

In this context radar imagery can be of immense value, in the first instance it may well be the only method by which imagery of the area can be acquired. This is certainly true of areas in South America, in Indonesia and even in parts of Africa such as Nigeria, where the acquisition of normal air photography is virtually impossible and even given the regular over pass of Landsat the possibility of receiving cloud free images is small.

Under such circumstances the apparent high cost of an airborne radar survey is well justified, it was shown in Nigeria for example that the acquisition of radar imagery for the whole country was potentially higher than a comparable air survey. In reality data were obtained in five months for the whole country whereas air photography, undertaken at the same time, was incomplete after ten years. Apart from the obvious costs of lack of data, inflation means that costs will inevitably rise over the ten year period so that the cost of air photography outstrips the costs of radar. For the future it is hoped that satellite radars designed for the needs of earth scientists, will yield valuable and regular data.

The second type of survey, the management of known plantations, is more liable to occur in the more temperate region such as Europe or N America. In these areas much of the natural forest has been exploited and lost, to be replaced with extensive plantations of commercial timber managed by a government forest agency and where controlled exploitation takes place. These forest areas are usually well mapped, the species mix and the age of growth are known. The areas involved can however be extensive and the management problems range from monitoring for crop damage, which can include pest infestation and the effects of acid rain, to illegal exploitation of timber and encroachment by agriculture into forest reserves. Also of importance is the assessment of a disaster as in a major fire or the effects of volcanic eruption on forest as in Mount St. Helens.

The difference between the two types of study is primarily a question of scale, the broad overview of the reconnaissance type survey compared with the more detailed view. The ability of radar to satisfy these surveys reflects the state of development of the SLARS, they have a proven value in the broader survey of forests but are only now beginning to achieve the resolution required for the more detailed surveys.

9.3   MAPPING UNITS AND INTERPRETATION

One of the major problems is deciding upon the mapping units to be used when interpreting forestry from radar, particularly in the tropical forest regional mapping. The nature of the backscatter signature from radar precludes identifying individual crowns and even if this were possible experience with air photographs points to the almost impossible task of identifying species from a crown in the mixed tropical forest. If the crown cannot be seen then there is no possibility of making estimates on tree height, bole size etc.

The radar image is composed of backscatter responses and shadow areas, in Figure 9.1 the relationship between the target and the image signature is shown for a forested area, note in particular how a scrub clump in savanna country could produce a signature that would be confused with that of a clump of taller trees. The backscatter is from the front surface and the

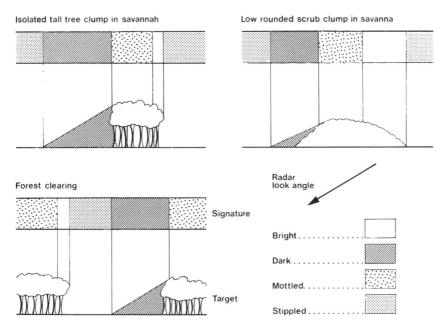

Fig. 9.1 The relationship between the radar target and the image signature for forest interpretation.

combined crown mass as a roughness surface, individual trees become part of this total surface.

At the outset of the survey in Colombia, S America, a claim was made that two mapping units would be identified, where one would contain trees 20–25 m high with heavy underbrush, 8 m spacing and bole diameters between 0.2–0.5 m, the second category would be 30–40 m high, continuous canopy, 13 m spacing and boles 0.5–1.0 m. Sicco-Smit commenting on these classifications, points out that this infers that it is possible to measure tree height from SLAR, to estimate understorey, count trees, and estimate bole size, in other words the volume of standing timber, this further points out that such a classification would be more applicable to forests in temperate zones than tropical rain forest. A subsequent study of large scale air photos and field visits found that there was no relationship between these forest types.

Experience in surveys such as that of the forestry of Nigeria showed that a division based upon species mix was more valid and that such divisions could be interpreted from other data or contextural characteristics. The basic division is made on tone and texture, identifying units of comparable tone and/or texture and labelling them with a code which may be further classified from associations but which should wherever possible be verified by field work. Tropical forests of great extent will cover hills and mountains with the result that any automated classification based on tone will be subject to variations due to the aspect to the imaging platform. Thus there will be areas of shadow or areas away from the direct view which will show a darker tone and areas on slopes directly facing the sensor which will result in a brighter return.

Under such circumstances the human interpreter has an advantage over the automated system by recognizing when differences are due to topographic changes and mentally compensating for these changes in order to continue to recognize a continuous mapping unit. Experiments are being conducted using a digital terrain model in order to calculate slope and aspect factors in advance. In reality these can only be of value in well developed countries where adequate topographic data exist. In the large forested regions there is rarely a reliable topographic map from which a digital terrain model can be constructed and even if stereoscopic cover were available only the slope of the canopy could be constructed and not the underlying terrain profile.

Experience of mapping forested areas has shown that trees growing in valleys or drainage channels will grow higher, because of the improved available moisture, than trees on hilltops where moisture and nutrients are in lower supply, the result can be an equalizing of the canopy in rolling terrain. This is particularly significant if interpretation is anticipated to indicate units of potential timber yield. The drainage pattern can be a significant indicator of forest types, for example drainage patterns can indicate low lying swamp areas, knowledge that such conditions exist indicates the probability of swamp forest which again can be linked to a definite species association. If the swamp is near the coast with a high probability of tidal influence then the immediate riverine forest could be mangrove.

SLAR images have been shown to represent topographic features better than optical imagery, this can be of value in forestry mapping. Not all the trees in a tropical forest are of commercial value; whilst the commercial timber can be dispersed throughout the forest, there can be a concentration related to growing conditions. These conditions can be shown to be related to topography, slope, solar aspect, potential soil thickness, drainability, coastal influence, altitude and similar criteria.

The human influence should not be ignored, the obvious elements here are the location of towns and villages and communication lines (roads, railways, tracks etc.). Towns and villages usually have associated agriculture and in many tropical regions this can be of the slash and burn type where an area is cleared of all timber and cleared by burning, enabling crops to be grown for 2–3 years. After this the area is left to regenerate as a secondary forest type and another area is cleared for agriculture. The pattern of such activity usually shows up very well on the SLAR image and can be identified as an area where the forest is under stress and there will be no commercial timber. Care has to be taken not to confuse cleared and burnt areas with standing water. In some areas more controlled cultivation takes place in which case it is necessary to identify where a forest reserve may be under stress.

Timber along roads has usually been exploited for commercial species with regenerated forest type resulting; excessive tracks or road activity in one area can indicate logging whether legal or illegal and this can usually be identified by SLAR.

Human activity usually results in a regular pattern so that plantations whether of tree crops such as rubber, oil palm or coconut, can be easily identified; in a similar manner areas of forest management are inferred.

A skilled interpreter with local knowledge and field data can make a reliable interpretation of radar imagery using these criteria, it has been commented that the final forest type map will be comparable to or even better in accuracy, costs and time than a map obtained by other remote sensing methods. As with all interpretive surveys, the final accuracy is related to the amount of field verification that is carried out.

9.4   THE VALUE OF RADAR FOR FORESTRY

It is interesting to record that forest mapping of tropical forests based upon the interpretation of radar imagery was undertaken as early as 1968 when a survey was carried out of the Darien area of Panama and Colombia. The final mapping identified four dryland forest units, four wetland forest units and three types of human influence associated with forestry. In this case the radar used was real aperture Westinghouse Ka-band.

The expense of radar surveys is frequently borne by a number of interested departments so that the same imagery may be used for geological interpretation land use, geomorphology and forestry. It is being increasingly recognized that a multi-disciplinary interpretation approach can improve each element of interpretation. For example, there is a growing appreciation of the relationship between vegetation and the underlying geology (Chapter 8). The Darien and subsequent projects in Colombia and Brazil were all multi-disciplinary and it is clear that forest classification for example can be improved by referring to the results of other interpreters.

9.5   PRO-RADAM – COLOMBIAN AMAZON (1973–1979)

The 'Projecto Radargrammetrico del Amazonas' in Colombia is an excellent example of the practical application of SAR to forest type mapping of a vast tropical rain forest area of which no accurate maps were available. The area of 38 000 000 hectares is covered mainly by dense natural rainforest with the exception of savannas and savanna forest types in the northern part.

The terrain consists of broad floodplain along the main rivers, a flat to low hill configuration of the dryland and in the centre a zone with higher hills and low mountains. The human influence of shifting cultivation, secondary forest and permanent grassland or agriculture is rather small in extent.

The project was multi-disciplinary incorporating forestry, geology and pedology. The total duration was 93 months and the amount of work was in total 6500 worker months.

In 1973 the area was flown with SLAR. The SAR system of Aeroservice Corporation in the X-band (3.12 cm) was used with a depression angle between 15°–39° and a flying height of 12 500 m. The flight line was N–S with always a west look direction. A total of 104 radar strips were flown with a 60% overlap having coverage in the near and the far range. The radar strip has a width of 37 km and a length up to more than 1000 km. The original scale is 1:400 000 with an enlargement to 1:200 000. A total of 69 semi-controlled radar mosaics at scale 1:200 000 were compiled. If cloud coverage permitted it, additional false colour aerial photographs at scale 1:80 000 were taken. The interpretation system is based on the differentation between wet and dryland types, the physiographic features like topography and drainage patterns and the differentiation between dense forest and the savanna types. The classification of the forest and vegetation types with the total area in percentage is given:

Region A: Forest and vegetation types of the floodplain (8.95%). Four sub-types and six combinations based on inundation characteristics.

Region B: Forest and vegetation types of the dry land (terraces and low hills).

(a) Dense and heterogeneous forest (69.20%). Six sub-types and six combinations based on topography.

(b) Savanna and savanna forest (15.05%). Five sub-types based on vegetation differences and topography.

Region C: Forest and vegetation types on high hills (3.85%).

Special Types: Human influences and non-forest (2.95%).

The 1:200 000 scale semi-controlled radar mosaics were used as base maps for the delineation and classification of the relevant forest and vegetation types. The final map was reduced to scale 1:500 000 (11 maps) and published in colour.

To verify the accuracy of the interpretation and to obtain information not visible on radar images of the species composition and its volume of the different forest types and vegetation, a reconnaissance forest inventory was carried out. The inventory consists of 874 one hectare sample plots mainly along the rivers (sampling density 0.0036%). For the statistical calculations the forest types are grouped according to the standing timber into the exploitable, with five sub-types, and the non exploitable, with two sub-types in the wet land, savannas, high hills and human influence. An overall map at scale 1 : 1 000 000 (two parts) gives the information of standing timber volume in combination with the ruggedness of the terrain. (La Amazonia Colombiana y sus Recoursos, 1979).

9.6   THE LAND USE AND VEGETATION SURVEY OF NIGERIA

In 1976 the Federal Government of Nigeria through the Federal Department of Forestry commissioned a radar survey of the whole of Nigeria with a requirement to produce a series of land use and vegetation maps of the whole country.

Nigeria can broadly be sub-divided into three main climatic zones: in the north the area is arid and semi-desert with few real forests but extensive cultivation; in the central zone the climate becomes more temperate with a woodland savanna country, some woodland and forest areas and extensive cultivation; in the south however it becomes humid and tropical with extensive wet tropical forest, localized cultivation and coastal swamps.

The decision to obtain radar imagery met with considerable criticism at the beginning, especially from the survey department. With the exception of the southern area which is under almost continuous cloud cover the country would seem well suited to air photography but Fig. 9.2 shows the air photographic cover at the date of the survey. In order for a land use map to have any significance it has to have an overall uniform time base and clearly this could not be satisfied with the existing photographic cover. At that time the Landsat cover for Nigeria offered no improvement, to the north the imagery was affected by dust storms and haze whilst in the south the cloud cover persisted.

The project was undertaken as two separate contracts, one for the acquisition of the imagery and the second for the data interpretation. The contract for image acquisition was obtained by Motorola (MARS) Ltd using their real aperture X-band system the modified APS/94 D. The system included a real time recording and display unit, simultaneous dual looking imagery and depression angle control. The flying commenced on the 2nd October 1976 and was completed on the 31st March 1977, a total of 181 calendar days of which 30 days were required for scheduled annual maintenance of the aircraft.

The total SLAR production time of 101 days included 146 daylight sorties and 14 night sorties. These sorties included 48% re-flights where imagery did not meet the required standards. The survey comprised 950 lines of flying to acquire the 642 planned flight lines, much of the re-flying was undertaken because of the effects of turbulence.

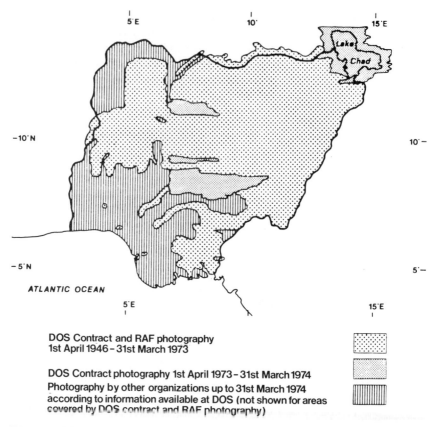

Fig. 9.2   The pattern of air photography acquisition at the time of the radar survey, Nigeria. (Source: Directorate of Overseas Surveys, United Kingdom.)

The radar imagery was obtained as two strips, relating to the two look directions involved. The strips were subsequently re-processed in the USA and compiled into 1.0 × 1.5 degree mosaics based upon the sheet layout and control points taken from the Joint Operation Graphics 1:250 000 scale series. Two mosaics for each sheet were produced, a 'North' look and a 'South' look (Fig. 5.2).

The interpretation team were supplied with copies of the strip data and a set of mosaic prints. The majority of the interpretation was carried out on the strip data and subsequently compiled onto the mosaics, this was found to be necessary since a certain amount of 'cosmetic' enhancement was performed on the mosaics in order to achieve a balanced result, to the interpreter this can mean a loss of valuable data.

The contract for interpretation was awarded to Hunting Technical

Services Ltd and the schematic programme of operation is shown in Fig. 6.1. The initial phase demonstrates how in any interpretation the inclusion of all other data sources forms an essential reference base. In the case of the Nigeria study all other available imagery was used, including Landsat, and extensive literature searches were conducted in order to help establish the vegetation keys to be used. The interpretation team then carried out a broad field study of Nigeria in order to acquaint themselves with the main vegetation patterns and the agricultural systems. During the field work, notes were supported by numerous ground photographs and these became invaluable in assisting interpretation. In this study the field teams consisted of foresters, soil surveyors, agriculturalists and land classifiers all with substantial field and interpretation experience.

During the second phase of the study the teams made an initial interpretation of the units based on tone and texture and Fig. 6.9 is a diagram of the relationship of these criteria to major units, in this instance contexture has been sub-divided into pattern, shape, size and association, where the latter refers to the relationship a unit may have to other features which help to determine its classification, e.g. mangrove forest will be found along river edges in the swamp of delta areas. Figure 9.3 is the same units shown as interpretation keys.

At this stage putting labels on the interpreted units was difficult, some areas could be classified on the basis of the previous studies and field work but in order to prove the classification and identify the constituent vegetation species further field work was essential.

Figure 9.4 is a series of typical radar units with their classification and some ground photographs of the area. Figure 9.5 is of an area in the north of the country; about one third of the way up the mosaic the strip appears darker than the rest, on the original strips this was more apparent, to the extent that detail in that region appears to be a negative of the same detail on adjacent sheets. This phenomenon caused considerable confusion during the interpretation until it was discovered from the flight log that the strip had been flown a month or so later than the adjacent strips. The aircraft had been restricted in flying due to seasonal dust storms, these storms herald the start of the rain thus the strip in question showed completely different soil moisture characteristics to its neighbours. This is an excellent demonstration of the problems that can occur in radar interpretation if all the details of data acquisition and the effect of various phenomena on the subsequent image are not known.

It was found that the more arid areas presented greater interpretation difficulties than the high forest region mainly because the wooded grassland was so sparsely wooded that the main signal response was from the soil, thereby giving the same signature as grassland. It was also noted that in the north villages could not be distinguished; this appears to be due to the

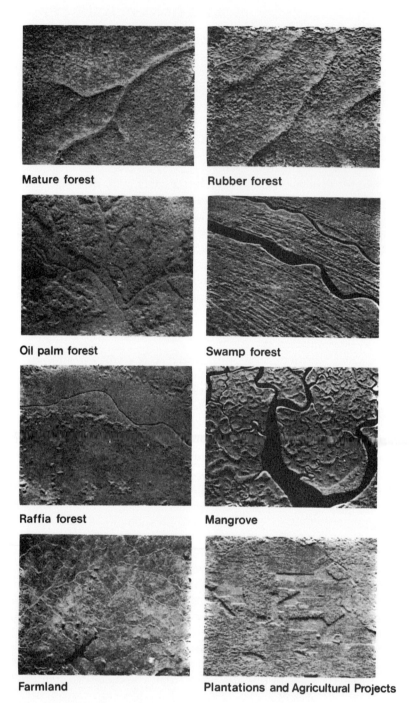

Fig. 9.3 Sub-formation SLAR signatures, Southern Nigeria.

Mature Forest, Southern Nigeria – Forest Zone

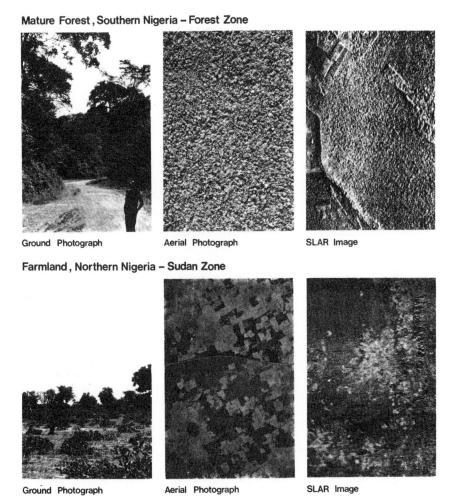

Ground Photograph   Aerial Photograph   SLAR Image

Farmland, Northern Nigeria – Sudan Zone

Ground Photograph   Aerial Photograph   SLAR Image

Fig. 9.4 SLAR photo key.

villages being constructed of mud huts with grass roofs which give the same response as the surrounding grassland. In the south the problem does not occur, the high rainfall makes grass roofs impractical and corrugated iron is used, this is an excellent radar target and gives a bright response. Figure 9.6a is an example of an interpreted image showing the initial interpretation prior to field checking, Figure 9.6b is the opposing look direction.

On completing the first stage of the interpretation the teams returned to carry out field checking, it is to be noted that the field teams were not a separate unit from the interpreters but that skilled field personnel were trained in recognizing radar images, having already had experience in

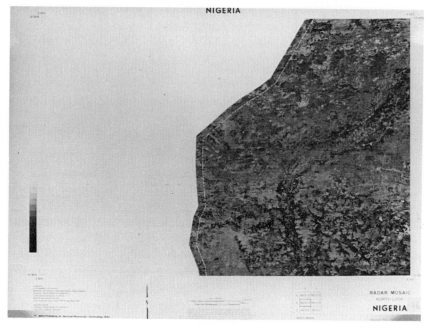

Fig. 9.5   Radar mosaic with multi-date imagery.

photo-interpretation. Such a system has a number of advantages, the interpreters can bring their practical experience to interpretation, essential when making interpretation based upon contextural data, and by returning to the field the relationship of the image to field expression is studied and becomes an iterative learning process.

During interpretation the teams could identify areas requiring identification or clarification and key type areas in a systematic mode so that subsequent field routes were already designated to the best economic and scientific advantage.

Detailed field observations were made of a wide variety of parameters as can be seen by the specimen field record pages given as Figs 6.2 and 6.3. An important aspect was the definition of key vegetation units and typical vegetation profiles have been included (Figs 9.7 to 9.9) to show how the unit species composition was obtained.

In the final stage the interpretation was adjusted or changed in line with the field observations and the units received their final classification. The interpreted overlays were then processed by cartographers to fit the base maps to be used and drawn into separations for colour printing.

The wide variety of vegetation zones meant that it was impractical to produce one legend for all the map sheets, instead each map sheet has in its

(a)

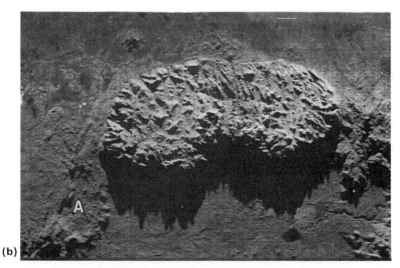

(b)

Fig. 9.6 (a) part of a Nigerian radar image showing the initial interpretation.
Key: A2   Undifferentiated grassland   H21a >60% cultivation
     A5   Shrub grassland   H21b 30–60% cultivation
     A6   Wooded shrub grassland   D10   Woodland
     J25 Forestry plantation   H37   Mixture of H21b and D10
     J30 Livestock project
(b) The same area as Fig. 9.6a in south look. Note how structures can be seen at A
which were not obvious on the north look.

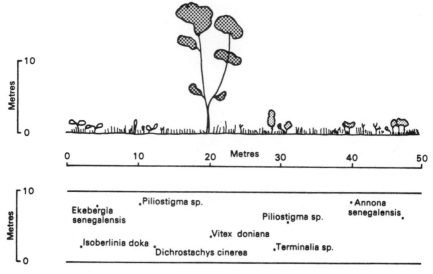

Fig. 9.7 Typical savanna vegetation profile.

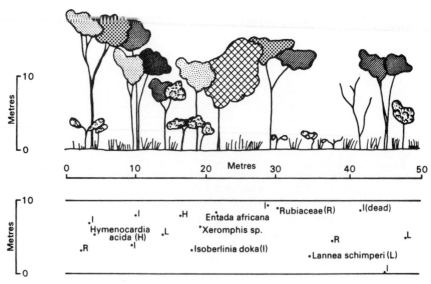

Fig. 9.8 Wooded savanna profile.

Plot size 113m x 6m

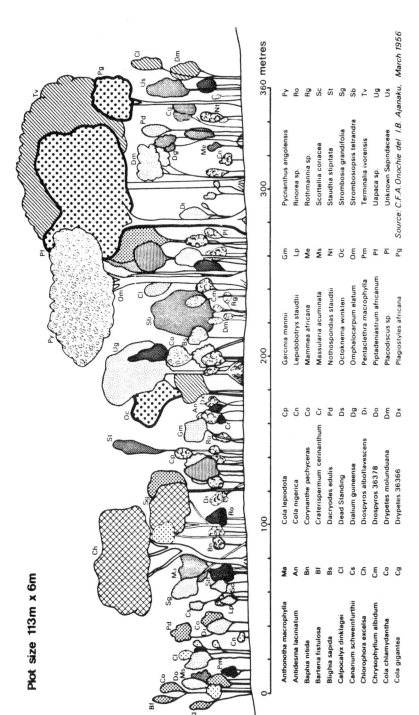

Anthonotha macrophylla	**Ma**	Cola lepiodota	Cp	Garcinia mannii	Gm
Antidesma laciniatum	An	Cola nigerica	Cn	Lepidobotrys staudtii	Lp
Baphia nitida	Bn	Corynanthe pachyceras	Co	Mammea africana	Me
Bartena fistulosa	Bf	Craterispermum cerinanthum	Cr	Massularia acuminata	Ms
Blighia sapida	Bs	Dacryodes edulis	Pd	Nothospondias staudtii	Nt
Calpocalyx dinklagei	Cl	Dead Standing	Ds	Octoknema winkleri	Oc
Canarium schweinfurthii	Cs	Dialium guineense	Dg	Omphalocarpum elatum	Om
Chlorophora excelsa	Ch	Diospyros alboflavescens	Di	Pentaclethra macrophylla	Pm
Chrysophyllum albidum	Cm	Diospyros 36378	Do	Piptadeniastrum africanum	Pt
Cola chlamydantha	Co	Drypetes molunduana	Dm	Placodiscus sp.	Pl
Cola gigantea	Cg	Drypetes 36366	Dx	Plagiostyles africana	Pg

Pycnanthus angolensis	Py
Rinorea sp.	Ro
Rothmannia sp.	Rg
Scottellia coriacea	Sc
Staudtia stipitata	St
Strombosia grandifolia	Sg
Strombosiopsis tetrandra	Sb
Terminalia ivorensis	Tv
Uapaca sp.	Ug
Unknown Sapindaceae	Us

Source: C.F.A. Onochie del I.B. Ajanaku. March 1956

Fig. 9.9 Undisturbed natural forest vegetation profile.

legend only those units appearing on the sheet, each with a comprehensive list of vegetation to be found. To achieve economies in printing, undertaken in the Nigerian Forestry Department's own unit, colours were used for the key units with sub-units indicated by a screen overprint. At the base of each map sheet smaller maps are shown: the geomorphology, the administration units and existing forest reserves, respectively.

The project ended with the production of a report which was unique; the first volume being a slim executive summary and the second main volume constructed as a training manual for introducing readers to radar so that other departments using the radar mosaics would have a better understanding of the techniques. The last volume was appendices containing field details and a report on the use of digital analysis.

This latter was a special phase of the project, Motorola SLAR was not recorded digitally but the Nigerian Forestry Department wished to determine whether digital analysis was possible since they had just invested in a digital image analyser for Landsat. A study was made of segments of images, which had to be digitized from film copies using a flying spot scanner. This part of the project had limited success but it helped to highlight the problems of digital analysis using radar images.

In the first instance imagery of this type, i.e. one waveband one polarization, results in a single monochrome image, therefore in analysis for interpretation there can be no subtle measuring of signatures on different images to determine a unique multi-spectral signature, only a basic density slicing. Density slicing is usually achieved by applying a distinct colour to units having the same tonal level, thereby making the distinguishing of units clearer and more accurate.

The radar image does present units which have apparent differences in tonal levels, but we saw earlier how the radar return presents a mottled or speckled image and not a continuous grey tone. There is as a result something of each tone in every unit, to varying degrees, largely in the form of bright spots and dark spots. It is the relationship of light to dark that gives rise to a tonal level rather in the same way as observation of a printed picture under magnification will reveal that the picture is not of continuous tone but is made up of numerous black dots.

Digital density slicing using computer analysis tends to see the grey levels in strict measured terms so that the resulting image can be more confused than the original.

Some experimentation revealed that by analysis of the image composition as the relationship or ratio of speckle to its neighbours, an improved result could be obtained; but in the final analysis it was considered that to use such techniques over the whole area would be extremely demanding in terms of computer effort and produce a less confident result than the use of human interpreters. It was clear that the human interpreter is able to absorb a certain amount of speckle and see an overall tonal unit, much as the observer does not recognize that photographs are reproduced as dots even when the image is greatly enlarged, a factor that has been used to advantage in some forms of modern pop art.

The Nigerian survey is unique despite the obvious size of such projects as Radam or subsequent surveys of other areas. It was the first time radar was used of a whole country to produce a definitive series of national maps combining field work with interpretation and including local training. The report as a training manual has already been described but throughout the project local staff participated in the study and at the end a short national training course was held.

At the end of the project 69 map sheets had been completed showing the land use and vegetation of the country which, including image acquisition, had occupied only two years.

When the final figures were analysed and it was realized that out of the original departmental budget not only had they obtained the survey but had built up a printing department, printed the maps, trained personnel and purchased major buildings and equipment, it was agreed that the same mapping results could not have been obtained by any other system without at least doubling the price which worked out to be US $6 per km^2.

9.7 NICARAGUA AND BRAZIL

Reference was made in the section on geological studies to the Nicaraguan survey and the Brazilian Radam survey. Both of these surveys included forestry as a part of their interpretation overlays and in a country such as Brazil mapping the vegetation is mapping a primary resource.

On the Radam project mapping was based on photo-interpretation of radar imagery, complemented with reconnaissance flights at low altitudes and ground observations, which enabled a physiognomic and ecological classification of the savanna (cerrado) and steppe (caatinga) vegetations.

The savanna covers the Maranhão–Piauí Paleo–Mesozoic sedimentary basin, and has three physiognomical aspects: parkland (parque), of the Tocantins depression; woodland savanna (cerradão), of the Maranhão–Piauí high plateaus; and isolated trees savanna (campo cerrado), of the São Francisco plateaus.

The steppe on the other hand occurs on levelled off metamorphic areas with sandstone remnants, presents an arboreal physiognomy in the Crateus-Parnagua peripheral depression, and a scrub vegetation in the São Francisco depression, both of them thorn bearing and deciduous.

Between the savanna and the steppe there is a contact area where savanna species grow together mixed with other species of the steppe.

Figure 9.10 shows a vegetation overlay prepared for one of the mosaics.

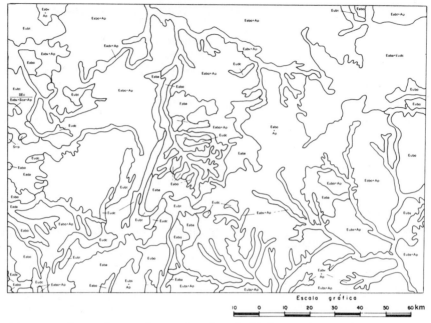

Escala gráfica

10   0   10   20   30   40   50   60 km

Fig. 9.10 Vegetation overlay produced on the Radam project.

CERRADO (SAVANNA)	Cerradão (Woodland Savanna)	**Sca**	em relevo acidentado – testemunhos
	Campo cerrado (Isolated tree Savanna)	**Srrp**	em relevo tabular
	Contato (Contact)	$\dfrac{\text{Sec}}{\text{grupos}}$	(encrave)
CAATINGA (STEPPE)	Caatinga densa arbórea (Closed thorntree Steppe)	**Eadv** **Eada**	em depressões em relevo acidentado
	Caatinga aberta arbórea (Open thorntree Steppe)	**Eaba** **Eabv** **Eabo**	em relevo acidentado em depressão longitudinal em relevo ondulado
	Caatinga densa arbustiva (Closed thornbush Steppe)	**Eudt** **Eudc** **Eudv**	em relevo tabular em relevo de cristas em vales e depressões
	Caatinga aberta arbustiva (Open thornbush Steppe)	**Eubt** **Eubm** **Eubs** **Eubo**	em relevo tabular – platôs em relevo tabular – mesas em relevo suave em relevo ondulado
	Ap(Agropecuária)		

This is one of the areas containing both savanna and steppe and the intermediate mixed zone (Projeto Radam, 1973).

9.8 DETAILED MAPPING

The preceding comments and description refer to the more extensive forestry survey, if we now turn to the other category of survey the map scale increases and with it the level of information required.

The requirements of a national survey in temperate areas with well managed forest reserves has been described in detail by Thallan and Horne of the UK Forestry Commission. The Forest Authority seeks data on the extent and composition of all woodland, irrespective of ownership, and this interest extends to non-woodland trees. The Authority has a function to perform to manage its own reserves and plantations but is also responsible for maintaining an overview of the total available trees in the country where exploitation or loss of woodland is vital to the overall national ecology and even the loss of significant single trees has to be recorded.

Normally a national census is conducted every 15 years, the requirement is for a system which will enable monitoring to be carried out at more regular intervals. The regular survey of the Authority's forests is considered essential for management and to provide locational information on land use and growth potential for budget planning, production forecasts, valuation and work programmes. It is also becoming common practice to maintain these data on a computerized information bank and any system has to be capable of providing data in a computer compatible format. There is also a requirement for more detailed localized data such as checking crop growth stocking and mapping natural catastrophies such as the effects of windblow.

Very few surveys using radar have been undertaken for these purposes and most of the work so far has tended to concentrate on specific research projects. Even in this respect very little work has been undertaken to ascertain the form of the interaction of microwaves with woodland or to analyse the ability of SAR to provide data on detailed species, age, yield or crop state.

Studies can be divided into two sections: system parameters and ground parameters.

9.8.1 System parameters

System parameters are those parameters which are considered optimal for the application to forest requirements, for example wavelength and polarization. A study of the literature shows that there is no consensus of opinion on the optimum wavelength for forestry. The Canadian SAR-580 campaign tended to indicate that L-band was preferable to X-band (Pala,

1980), the L-band being found suitable for differentiating three levels of regeneration, however Goodenough (1980) considered X-HH best for visual analysis. Work in Arkansas with Ka-, X- and L-band suggested the L-band gave a reduced dynamic range and many researchers express a preference for shorter wavelengths. There is a growing realization that perhaps forestry's needs are best served by a multi-frequency approach using the data in combination.

In polarization the use of like polarized data is shown to be preferable, but there is no distinct indication as to whether this should be HH or VV; there is some evidence (Peterson, 1968) that cross polarization helps define forest/non-forest boundaries.

Resolution is at present seen to be preferable at between 6 and 10 m but evidence is inconclusive. Similarly with look direction and incidence angle there is too little real research work to come to any conclusions. In areas of rugged terrain more than one look direction is preferred if only to remove some of the effects of aspect.

Multi-temporal imagery is seen to have advantages; this is obvious when dealing with deciduous woodland when species can be identified by leafing periods. There are also advantages when trying to monitor infestation, but on the whole, because woodland can be seen as a long term crop, multi-temporal imagery is less essential than for agricultural requirements.

9.8.2   Target parameters

Of the ground parameters the most fundamental is the delineation of woodland areas from non-woodland. In general this has been achieved but under certain circumstances anomalies do occur. Study of Seasat imagery showed that woodland could be reliably mapped (Hunting Geology and Geophysics Ltd, 1981) but there have been examples where this differentiation has not been possible as reported by Ulaby and Batlivala (1980) when woodland could not be differentiated from corn on L-band radar. Species identification as a parameter has been succesful in delineating broad species categories as far back as 1967; using the Westinghouse Ka-band radar it was possible to discern pine forest, juniper woodland, grassland and sage brush, a later study using the same radar subdivided ponderossa pine, juniper woodland, white fir forest, hardwood forest, sage brush, shrub, grassland and recent burns (Peterson et al., 1969). Despite this and other successes the delineation of species remains an area which has not been substantially researched, but recent indications are that it would be extremely rewarding.

As part of the SAR-580 campaign in Europe where a number of test areas were flown using a multi-frequency SAR, a test area over the UK Forestry Commission's major forest in Thetford, Norfolk was studied (Churchill and

Keech, 1983, 1984, 1985). The test site was imaged in X- and C-bands and, as was common with all experiments carried out in this campaign, the quality of the imagery was not as high as had been expected.

Examination of the optical imagery, especially the X-band, showed that Corsican Pine and Scots Pine could be separated. Subsequently efforts were made to combine the imagery in order to produce colour composites. It is interesting to note that the researchers did not have ready access to sophisticated equipment; to test the potential of colour combinations a very simple system was used, 70 mm slides were made of the imagery and three projectors used to project the images through colour filters and to superimpose the projection. Simple though this system may sound, it demonstrated that a higher level of interpretability could be achieved by combination and that the interpretation could be extended to include mixed woodland and to differentiate plantations of different age as well as species (Plates 1 and 2). Subsequent experiments used digital techniques and these confirmed the original findings. The researchers noted a correlation between tone and texture and age which appeared to be associated with the thinning practices used giving rise to a decreased stand density with a resultant roughening of the canopy. Other researchers using L-band SAR in HH and HV mode could classify Maritime Pine (Pinus Pinaster) into three age groups 0–4 years, 4–10 years and > 10 years (Le Toan et al., 1980).

In detailed forest mapping it is desirable to be able to obtain data on tree geometry, i.e. height, crown size, bole size etc., this has yet to be satisfactorily achieved by large scale air photography or other remote sensing means so any comments on the ability of SAR have to be seen against this background.

In general these characteristics cannot be identified on SAR images, it may be possible to infer tree height from radar shadow data and it has been shown that leaf orientation and shape can affect the backscatterers return so that the tone and texture of tree canopy can be related to leaf factors which can in turn relate to crop maturity. In some cases the length of the pine needles can be comparable to the radar wavelength and act as good resonators thereby give a high signal return (Dellwig et al., 1978; Schuchman and Lowry, 1977).

Without data on tree size it is very difficult to make statements on potential yield except where yield can be related to crop state. This latter involves the assessment of damage by infestation in a tree crop. In the studies referred to earlier (Le Toan et al., 1980) not only were age groups defined but also two density classes. Churchill noted in his SAR-580 experiment that the effects of the root disease Fomes Annossus could be detected as could the results of windblow. Given that it is possible to recognize species in plantations and relate the signature to crop age and density, it is possible to use known potential yield data to make a yield assessment. Such a calculation could also take into account the potential effects of damage or infestation.

In tropical forests, given the species composition of a forest type a similar estimate of potential commercial timber is conceivable for each identified unit, sample field data would help to confirm these estimates.

It has been assumed that the backscatter response has been derived from the canopy, this cannot be verified, it is known that radar can penetrate certain materials and it is reasonable to assume that they penetrate vegetation to cause some of the return to be internal, i.e. below the surface canopy. Various attempts have been made to emplace corner reflectors at different levels in a forest, but so far no satisfactory results have been obtained, certainly none which could specify the amount of penetration for different species.

9.9   CONCLUSION

In detailed forest surveys there is evidence to show that species and age identification are feasible and that monitoring crop state is definitely possible. In reality the possibility of obtaining regular airborne cover for these purposes must be largely discounted, it is most unlikely that a forestry authority will receive funds for regular flying of radar, most have difficulty in obtaining intermittent air photographs.

The potential for satellite radar is different, regular cover with different look angles could provide vital data for the efficient management of many forest or woodland areas for national authorities.

There is little doubt that for the larger inventory surveys, radar can be demonstrated to be a real and valuable tool. Furthermore, recent research has shown that where radar is combined with multi-spectral imagery, the addition of texture from the radar enables otherwise seemingly uniform tree cover to be sub-divided into species groups, thus SAR is adding a dimension that is not possible from another source.

REFERENCES

Churchill, P. and Keech, M. (1983) European SAR-580 Campaign Investigators' Interim Report.

Churchill, P. and Keech, M. (1984) *Multifrequency analysis of SAR-580 Imagery for Woodland Determined in Thetford Forest, England.*

Churchill, P. and Keech, M. (1985) European SAR-580 Campaign Investigators' Final Report.

Dellwig, L. F., Bare, J. E. and Gelnett, R. (1978) *SLAR for clear as well as cloudy weather.* Proc. of the international symposium on Remote Sensing for Observation and Inventory of Earth Resources and the Endangered Enviroment, Frieberg, R. R. Germany, III, pp. 1527–46.

Goodenough, D. G. (1980) *Quantitative analysis of SAR data for detecting forest infestations of insects.* Final Report of the Airborne SAR Project, Intera. Env. Consultants Ltd, Report ASP-80-1.

Hunting Geology and Geophysics Ltd (1981) *Final report of the data content of overland Seasat-A SAR imagery.* Report to RAE Farnborough.

La Amazona Colombiana y sus Recousos. Proyecto Radargrammetrico del Amazons, Bogata (1979) (Consultants' Report).

Le Toan, T., Shahin, A. and Riom, J. (1980) *Application of digitised radar images to pine forest inventory: first results.* Proc. 14 International symposium on remote sensing of the environment, San Jose, Costa Rica, 2, pp. 943–4.

Pala, S. (1980) *Radar imagery for assessment of coniferous forest regeneration.* Final report of the airborne SAR project, Intera Env. Consultants Ltd, Report ASP-80-1.

Peterson, R. M. (1968) *Observation in the geomorphology and land use of part of the Wasatil Range, Utah,* US Geo. Survey Interagency Report, NASA-140.

Peterson, R. M., Cochrane, G. R., Morain, S. A., Simonett, D. C. (1969) A multisensor study of plant communities at Horsefly Mtn., Oregon. In *Remote Sensors in Ecology* (Ed. Johnson) Georgia University Press, pp. 63–93.

Projecto Radam (1973) Vol. 1, Final Report to the Government.

Schuchman, R. A. and Lowry, R. T. (1977) *Vegetation classification with digital X-band and L-band dual polarised SAR imagery.* 4th Canadian symposium on Remote sensing, Quebec.

Ulaby, F. T. and Batlivala, P. P. (1980) Crop identification with L-band radar. *Photogram. Eng. and Remote Sensing,* **46**, (1), 101–5.

10

Land use

10.1 INTRODUCTION

The term 'land use' is seen to cover both rural and urban applications and in this context will be taken to include crop identification. Much more research has been carried out into the detailed analysis of the interaction of microwaves with crops than has been applied to forestry and woodland and in this instance more general vegetation can be included. Like forestry, land use can be seen as being applied to broad regional surveys, it can also be applied to detailed crop investigations.

The Nigerian survey referred to under forestry was also titled a land use survey. In a country that is not entirely covered by forest, the remaining land has to be classified. This is especially true if that land carries occasional trees which can be considered part of the overall timber resource as in Nigeria. The Nigerian survey is therefore an excellent example of the practical application of SLAR to land use mapping, except that no attempt was made at urban classification.

Other projects described elsewhere have also included an element of land use, the South American Radam and Proradam studies included interpretation of soils, geomorphology and general vegetation to produce a map of the existing land use, similarly the Nigerian survey resulted in a land use map.

There is ample evidence to show that radar is suitable for the production of regional land use maps, however away from the high rainfall areas there is real competition from visible band satellite data. It is anticipated that when satellite radar data become regularly available it can be combined with satellite MSS data to extend the interpretative potential of the latter in the same way as has been evidenced by localized experimentation referred to earlier.

10.2   THE REQUIREMENTS FOR LAND USE

The requirements from a radar system for larger scale land use mapping have not been satisfactorily stated, but a number of parameters have to be considered. For a land use map to be of any significance it is essential that all the data are at a comparable date line. This was one of the main reasons why the Nigerian survey was designed to employ SLAR, to use air photographs which could be spread over several years would yield a map with non-coherent data content. At least using SLAR all the Nigerian imagery was within a five month time span, something air photography could not achieve. Even with Landsat this is not always possible, although closer datelines are achieved. Given an 18 day repeat cycle an area can be covered in a very short time span, provided no cloud occurs, unfortunately weather conditions in many areas can alter very quickly such that to obtain cloud free cover of a temperate region can be very difficult indeed. For example, in 1984 the UK experienced one of the best summers on record, with what seemed like months of near perfect weather leading to drought conditions. In reality full cover using the Landsat Thematic Mapper was not possible for that period, mainly because of constraints of severe haze along coastal regions. If there is to be a reliable regular imaging system for regular land use mapping, then radar would seem to be best suited to this role.

Land use, unlike geology, is seasonally dynamic and indeed is more changing than forestry. Most requirements are not only for maps of existing land use but also for a system to regularly monitor change. These changes can be urban expansion and the loss of agricultural land, changes in river regimes, the effects of shifting cultivation, the spread of erosion and desertification and so on. This therefore requires not only the identification of features on the radar but the comparison of subsequent data in order to recognize when valid change has taken place.

In cropped areas there is a requirement to identify crops, to monitor their growth and health and to use the data in order to predict potential output or yield. In regions largely of monoculture this is not a major problem, areas such as the large wheat growing belts will only need to identify types of wheat. In many farming areas the cropping pattern is much more complex, with crops varying from the major crops of wheat, sugar beet, potatoes and so on, to local market crops and mixed vegetables or fruit. This in fields that vary considerably in size, that are on widely different soils, that may have field boundaries of hedges or stone walls or none at all, that may vary in management procedure and that may be on hill areas presenting a varying aspect to the sensor.

Of course these factors apply whatever the sensor, and thus optical sensors including air photography have to overcome the same problems. In assessing the success of microwave systems in differentiating crops it is necessary to question whether optical systems are reliable and have had a better success rate. In determining a spectral signature for a crop the analyst is assuming that all crops of the same type have the same spectral appearance at each stage of growth. Unfortunately the crops are affected by the soils on which they grow, the water retention potential of that soil, in conjunction with slope and aspect and the consistency of farming methods. This leaves aside the effects of colour changes due to meteorological conditions, sun angle, atmospheric attenuation and so on. The result is that no reliable spectral signature key has yet been devised, other than for local cropping, and the reliability of interpretation of MSS data is related to the amount of field work supporting the interpretation.

The problems confronting microwaves have some similarities, for example the backscatter from vegetation is a measure of the dielectric constant or water content of the plant. This in turn is a measure of the moisture capacity of the soil. The backscatter signature of a plant can be shown to be a factor of the structure of that plant, the size, density and angle of its leaves for example. The ability of a plant to demonstrate an ideal structure is, in part, a measure of the climatic conditions available for that plant's growth and also a reflection of the farming methods used.

It is unlikely that a totally accurate identification method will evolve from either system, it therefore becomes a question of which system offers the most promise.

10.3   MODELS

In an effort to understand the elements that affect the backscatter coefficient from vegetation a considerable amount of research effort has gone into constructing theoretical models for the interaction of microwaves on vegetation and with the soil surface.

In considering the models for soil surface these can be broadly divided into three groups;

(a) Small perturbation model
(b) Kirchhoff's approximation
(c) Two scale models.

with the modelling to the soil dielectric as the prime requisite for all models.

The objective is to construct a mathematical model that will exhibit the properties of different surface types that can be said to be found in soils. The small perturbation model (Ulaby et al., 1982) describes a surface with minor irregularities on an otherwise planar surface.

Kirchhoff's theory is based on the assumption that the surface can be modelled locally by a series of tangent planes and so a plane boundary reflection occurs at every point on the surface. To satisfy this assumption the surfaces must be smooth and have a large radius of curvature thus presenting an undulating surface.

The two scale surface is a combination of the first two models in which a small scale perturbation is superimposed on a large scale undulating surface (Figure 10.1). The calculations then examine the forms of the backscatter equation for a range of surface parameters over a variety of incidence angles using the various microwave wavelengths.

It is difficult to decide which of the models available to apply, a soil surface in practice will not have single-scale variations in profile but will be many-scaled.

When the interaction of microwaves with vegetation or plant cover is considered then there are similar models and in this case four categories are usually identified.

(a) Empirically derived models
(b) Dielectric slab models
(c) Lossy scatterer models
(d) Random media models

10.3.1 Empirically derived models

As the name implies, these models have no theoretical basis but are derived using experimental data.

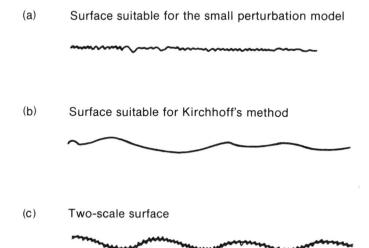

(a)   Surface suitable for the small perturbation model

(b)   Surface suitable for Kirchhoff's method

(c)   Two-scale surface

Fig. 10.1 Scattering surfaces.

10.3.2   Dielectric slab models

Here the vegetation layer is regarded as a dielectric slab lying on the soil surface where the slab's dielectric constant is considered to be related to plant height, density, moisture content and other plant properties. The backscatter can then be considered as the sum of a scattering term from the canopy, and a term from the soil which has been suitably attenuated by the slab.

One of the first such models was developed by Bush et al. (Bush and Ulaby, 1976) and a subsequent model was developed by Attema (Attema and Ulaby, 1978).

Attema modelled the vegetation as a water cloud. The dielectric constants of air and dry vegetation are approximately 1.0 and 1.5 respectively, whereas the dielectric constant of free water is considerably higher. This was used as the justification for considering the plant canopy as a cloud of randomly distributed water droplets. The model has been further developed (Hoekman et al., 1982) to introduce angular dependence into A and C which is particularly suitable for crops such as potatoes where leaf orientation is important.

The various models that have been developed have been tested against scatterometer data with the experimental data.

10.3.3   Lossy scatterer

In these models the plant foliage is regarded as a series of regular geometric shapes, lossy scatterers, using the shapes and their distribution to simulate plant geometry

The model was originally developed to simulate the effects of trees over water (Engheta and Elachi, 1981). In this context trees were represented by a random distribution of spheres. In subsequent models the shapes considered were circular discs which were taken to represent plant foliage (Lang, 1981; Lang et al., 1982 and 1983). The models have been tested against experimental data using deciduous trees as a target with a reasonable degree of success.

10.3.4   Random media

This model was first described by Fung (Fung et al., 1977) where the model considered a half space filled with vegetation and was developed (Fung, 1979) to consider a slab of vegetation lying between the air and a perfectly flat soil surface.

The work on the interaction of microwaves with a wide variety of surfaces and vegetation canopies occupies a great deal of research. This very cursory background by no means does justice to the work that has been carried out; on the other hand the average interpreter need only be aware of the research that is taking place and will not necessarily be involved in the theories of interaction.

10.4 IMAGE INTERPRETATION 1

In considering the spectral signature of a cropped area obtained from observation by optical devices, a precise measurement of the return signal in grey level terms is made for each spectral channel. It is the unique combination of readings which provides the so called signature. This grey level reading is fairly easily determined when dealing with a continuous tone image, there may of course be variations in tone due to internal differences on a crop.

The microwave image produces an immediate problem in this respect, inspection of an image will reveal the major visual difference between the microwave image and the optical image. The continuous tone grey of the optical image is no longer available as the microwave image is made up of grey responses and a varying pattern of white points or speckle.

The identification of a unit by an interpreter is part recognition of a grey level given by the dark response areas and part acceptance of a degree of speckle and speckle pattern. This presents no great problem in visual interpretation as the observer can accept all the factors as presented and correlate two areas of comparable overall appearance. This leads many workers to consider visual interpretation of the single image to yield the best results (Kessler and Jano, 1985) especially when allied to field work.

As with optical image interpretation the long term objective is to develop methods of computer aided classification leading to a fully automated image analysis system. To achieve this there has to be a viable method of reading a signature or response for each distinctive unit that either removes the speckle or takes into account the grey scale value and the speckle pattern.

As with forestry it is necessary to examine the parameters that affect the representation of crops on an image and their subsequent interpretability, these again are divided into the parameters of the system itself and those presented by the crop, the so called ground parameters.

10.4.1 System parameters

(a) *Wavelength*
A large number of primary researchers studying the interaction of microwaves with plants use scatterometers for their experiments. These scatterometers are mounted onto mechanical arms (known as cherry

pickers) or on mobile towers. In these cases the angle of incidence can be varied and the platform moved either to different observation positions to give different look directions or moved continuously during recording. The latest tests have been conducted by mounting the scatterometer into an airborne platform, usually a helicopter.

Such tests have the advantages that all parameters can be controlled and measured, including soil structure, moisture, plant geometry, climatic condition and so on. Most of the data for the various parameters have been devised from such studies and few of the comments or results have been tested using imaging radars. The main experiments in this field have been those carried out by the team of the University of Kansas, formerly Ulaby, Moore and Fung and by the ROVE team working in the Netherlands and made up of representatives from research institutes and universities.

By using scatterometers, classification accuracies of the order of 90% have been achieved (Bush and Ulaby, 1978), where operating in a range between 13–16 GHz the best classification was obtained at 14.2 GHz with VV polarization for corn, soya bean, milo, wheat and alfalfa. C-band at 4.75 GHz has been found (Paris, 1983) to be optimum for delineating mature corn and soya beans, although some researchers found the L-band data had an advantage in recognizing the larger grains (Mehta, 1983). When examining the broad leaf crops such as sugar beet and potatoes the 9.4 GHz wavelength was found to be optimum (Hoogeboom, 1983) whilst for paddy rice, corn and cotton, (Pei-Yu, 1983) good results were obtained using the X-band.

It is therefore evident that no one frequency can be said to be preferable for crop discrimination and the frequency to be used does to some extent depend upon the crops to be mapped.

This is no different from findings with optical sensors where single frequency recording reduces the discrimination levels and it is only the introduction of multi-spectral imaging that has increased the interpretability of optical images. It is therefore reasonable to assume that the combination of frequencies to produce composite images would increase the potential for microwave analysis.

During the Canadian SAR-580 campaign experimenters found that using X- and L-band imagery in combination improved classification capability (Protz and Brisco, 1980; King, 1980 and Crown, 1980). Subsequently experience with the SAR-580 campaign endorsed these findings (Gombeer, 1983; Sieber and Trevett, 1983; Mégier et al., 1985); some of the results are discussed in detail later in this chapter.

(b) *Polarization*

Much less work has been carried out with different radar polarizations mainly because there are very few sensors developed that can make multi-polarized measurements. Certainly there are no space radar data at present in more than one polarization.

Early evidence (Bush and Ulaby, 1977; Ulaby, 1981; Le Toan, 1982) would tend to infer that cross polarized data offered no great advantages. Subsequent studies have shown that a combination of HH polarizations in different frequencies seen against VV polarization combination for the same frequencies can demonstrate significant differences on the image which improve classification (Sieber and Trevett, 1983).

In parallel with the studies on frequencies, evidence is increasing that changes in polarization provide a different view on an image of a scene which by itself may not yield greater interpretation benefits but when viewed as another spectral channel and viewed in combination with other frequencies or wavelengths, the interpretation capability increases with particular reference to the discrimination of crops or vegetation.

(c) *Resolution*
There is very little evidence of the optimum resolution required for crop discrimination and the role it plays. Researchers in general tend to seek an improved resolution quoting 3 m as desirable, some going down to 2 m.

It is clear that given a nominal resolution of between 10 and 20 m, the resolution does not play a large role in crop discrimination compared with frequency and polarization.

(d) *Angle of incidence*
Most imaging radars have a fixed angle of incidence and only the SIR-B had any facility for varying the angle and results from this experiment have still to be fully analysed.

As a result most of the data on the role of incidence angle are derived from scatterometer tests and have yet to be borne out by comparative studies of imagery.

What is evident is that the ability to discriminate some vegetation can be seen to be related to the incidence angle, thus Pei-yu (1983) considers the optimum angle for paddy rice to be between 42° and 72° whilst Paris (1982) considers that an incidence angle of 50° with C-band is best for delineating corn and soya bean; Shanmugan *et al.* (1983) found that for an angle of up to 40° the accuracy of crop classification improved but once past 40° it remained relatively constant.

There is a growing body of opinion that considers that multiple incident angle imaging would further improve interpretability. Given that SAR is viewed at an oblique angle that changes across an image some variation can be fairly easily achieved by increasing the lateral overlap between image strips thus providing views of the same area at different incidence angles (Fig. 10.2).

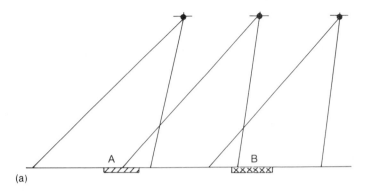

(a)

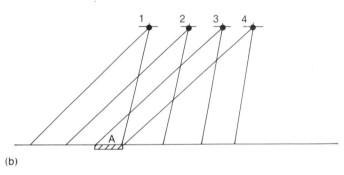

(b)

Fig. 10.2 Effect of increasing lateral overlap. (a) Flight paths arranged for economically obtaining full area cover, note how a field A is imaged in near range whilst a comparable field B is in far range, which could result in a difference in image representation and therefore subsequent interpretation. (b) Flight paths arranged to give increased lateral overlap. Field A originally only in near range, is now in near range view 1, mid-range view 2 and far range view 3; field B would also have had similar coverage giving comparable image responses. Note also that such a flight plan permits for stereoscopic viewing.

(e) *Multi-temporal imagery*
Vegetation, particularly crops, is subject to strong seasonal changes. A field shown fallow on one image can be supporting a major crop later in the same year and without this data land use classification will not be at maximum value. Similarly for crop yield prediction it is necessary to observe a crop throughout its growing stages, this monitoring is also essential to minimize the effects of infestation by providing advance warning to farmers.

As with any survey, regular imaging by airborne techniques can be expensive, however radar can be shown to have better economics in this respect than optical systems since flights can be arranged to an advance plan certain in the knowledge that, barring exceptional storm periods, images will be obtained.

The greatest potential for multi-temporal imaging must be with space borne systems and here the same advantages occur. A system such as Landsat may overfly once every 18 days but imagery will only be of value on clear days, the radar has no such constraints.

Where multi-temporal data have been available, improvements in interpretation have been demonstrated, for example a 15 day revisit period using dual polarized data enabled a 90% classification rate to be achieved for corn, milo, soya bean, wheat and alfalfa (Bush and Ulaby, 1978), in another study multi-data and C- and L-band radar was used to obtain an improvement from 85% to 100% in the classification of corn (Shanmungan et al., 1983).

10.4.2   Ground parameters

(a) *Crop type*
Clearly if the objective is to discriminate different crops and to identify them then the type of crops involved is of paramount importance. In the first instance it is essential to gather as much data about an area as is possible, the need for initial study and field work prior to interpretation is continually to be emphasized, these data should include a list of the types of crops that can be expected to occur.

It will be realized from the comments in systems parameters, that interpretation results have been obtained on most of the major crops but that there remain many crops throughout the world for which comparison data are not available. This is not to infer that they cannot be identified, careful study of radar images over an area, whether single frequency or composite of multi-frequency dual polarization data, may well indicate differences in crop units which local knowledge and subsequent field work will be able to classify into specific crop types.

(b) *Crop age*
Experiments have shown that the backscatter return from a plant will vary through the plant's life cycle and that, even at an early stage of growth, differences can be discerned. In an experiment under the European SAR-580 Campaign (Sieber and Trevett, 1983) it was shown that in a field containing two crops (half the field sown to peas the other half to french beans, both of which were about 10–15 cm high) could still demonstrate different signatures when observed as combined X- and L-band imagery in a colour composite, enabling the two segments of the field to be clearly

identified. Other experiments using multi-temporal data (DeLoor and Hoogeboom, 1982) could detect the differences due to the growing stages of sugar-beet, potatoes, spring wheat and oats. The crop can be observed and differentiated at various stages of its growth provided multi-temporal imagery is available.

(c) *Crop geometry*

The leaf angle of a plant will vary throughout the day, being conditioned by the sun angle and the moisture content. It can be demonstrated from the models using lossy scatterers that the backscatter return will change with leaf angle and subsequent studies with scatterometers have endorsed this finding. For comparative studies it is therefore ideal if imagery is carried out at the same time of day and under comparable conditions; since this is not always possible, information on local climatic conditions should be recorded.

The form and position of seeding heads can also be of significance, it has been shown (Sieber and Trevett, 1983) that two fields of wheat which demonstrated different backscatter returns were probably the results of plant structure. In one field the heads of the wheat were upright and therefore a continuing line of the stem, in the second field the wheat heads had bent over at ripening, and the size of the heads compared to the wavelength of the transmitted SAR.

In the case of vertical ears most of the incidence energy will be reflected away from the look direction and what is scattered back to the radar may be a multiple scattering process between different plants or the ears and the underlying rough soil surface. A bending ear illuminated by an incident radar wave has a bigger geometrical cross-section compared with a vertical ear, therefore the dielectric geometry of the dominating dielectric scatter would seem to explain the differences in the backscatter intensity. This phenomenon of a brighter backscatter return from the bending ears was only observed on the X- and C-band imagery but not on the L-band, it was determined that the L-band had the greater penetration so that the backscatter came largely from the soil surface.

A laboratory experiment has since been conducted by Ulaby in the USA whereby a number of plastic tubes, representing plant stalks, were first observed at a variety of frequencies. Copper collars of different lengths were then introduced over the tubes and measurements made with the collars at different heights at different frequencies. When the size of the collar equalled the size of the wavelength transmitted a strong resonance effect occurred even when the collar was not at the top of the artificial stalk.

In the same study a further effect due to plant geometry was observed: in one particular field there was a distinct pattern of banding which was too wide to be a factor of row direction. It subsequently transpired that the field

THETFORD FOREST CHV-RED XHH-GREEN XHV-BLUE

Plate 1 Digitally combined SAR images of the Thetford Forest, Norfolk, England. Data from the European SAR–580 campaign have been used; in this example CHV is in red, XHH is in green and XHV is in blue. The combination was used to identify different tree species and could be used to classify plantations on age differences. (Hunting Technical Services Ltd.)

Plate 2 The same scene as in Plate 1, this time a specially designed filtering routine has been used in an attempt to minimize speckle and to obtain units that could be automatically classified. This is part of ongoing research, but already the potential can be seen.

Plate 3 Combined mosaics forming a coloured image. In this case two X–bands (VV and HH) have been used in combination with C–VV. (Hunting Technical Services Ltd.)

Plate 4 Digital data combined with boundary information. In this case only two images have been combined, L–HH and X–HH, but the addition of the boundaries already clarifies the units. (Hunting Technical Services Ltd.)

had been harvested at the time of imaging using a system of three threshing machines operating in parallel, measurements showed the bandwidth to be equal to the measurement of the three harvesters which crossed a field, turned and returned parallel but in the opposite direction to the preceding cut. The result was a different lie of the stubble showing as banding on the image. This banding was observed on the VV image but not on the HH.

As a final form of crop geometry it is necessary to consider row direction and row spacing. The row direction can be shown to affect the backscatter return of a crop, the backscatter return can be higher by 10–20 dB when looking along the rows when using C-band HH imagery (Paris, 1982) and on L-band imagery the effect of row patterns has been found to equal or exceed the effects of soil moisture or vegetation differences (Waite *et al.*, 1978). Researchers on the European SAR-580 campaign, however, did not feel row direction made a significant difference to the signatures.

(d) *Crop state*
There is very little evidence of the detection of changes in crop state due to infestation, although the separation of healthy from blighted corn has been noted (Ulaby and Moore, 1973). This is another instance where multi-temporal imagery will be necessary, unless comparison can be made on an image to known areas of healthy crops.

(e) *Other factors*
The return from a crop has been shown to be related to the dielectric constant, therefore soil moisture must be of significance in delineating vegetation cover types (Shanmugan *et al.*, 1983). In a similar manner, especially with emerging crops or when using radars with frequencies such as L-band which have some penetration, the soil texture is a factor in the viewed return. There are other factors which could affect an image but there is no evidence available to determine whether this is so or by how much. In this category comes the difference between the daylight and the night time image, of significance when the independent illumination potential of radar is considered. Satellite radars will be able to image in both descending and ascending modes, one of which will be in daylight, the other at night.

10.5   IMAGE INTERPRETATION 2

The interpretation of single band black and white imagery is comparable to the interpretation of any monochrome image. In an agricultural area, especially in highly developed countries, the field patterns should be obvious and stand out clearly, differences in crop types can be distinguished by their tone/texture values.

As with all back interpretation there is no easy way of labelling the units, this will require local knowledge, field data and subsequent checking. The accuracy of the classification will be related to the ancillary data available.

In non-agricultural areas it is possible to use the same basis of tone/texture to delineate different vegetation cover zones, their classification may be helped by contextural information and local field data.

Urban areas normally demonstrate a high backscatter return with some evidence of pattern due to roads or other similar features. In some cases internal classification on a broad basis can be made, for example in an old city such as Freiburg in Germany the older city area can be differentiated from the area with houses and large gardens, the modern housing estates and as a final unit the area with new modern office blocks.

In the city of Neuf-Brisach, also in Germany, where the city dates back to the 1700s the strong geometric walls of the old city can clearly be recognized. In this instance not only were two river bridges identified but they were observed to be different and field checking showed the differences were related to their construction one being metallic the other concrete (Endlicher *et al.*, 1985).

At this stage it is worth recording that the hydrology on an image can be readily interpreted and that this in turn can help to relate to land use. This has already been noted as a factor in interpreting forestry and much the same criteria apply, especially when the drainage is seen to be artificial and relate to irrigation schemes.

10.5.1 Multi-specular radar

The greatest potential for interpretation must be in the use of combined frequencies, look angles, look directions or polarization, and to carry this out efficiently requires access to a computerized image analyser. Ideally the base data itself should be digital if the full value of the data is to be exploited, but digitizing optically correlated images will still yield improved results.

It is difficult to use the term multi-spectral as used with optical systems and in Landsat, and multi-frequency has neither the same significance nor does it allow for different viewing angles, multi-polarization or any of the other techniques. Since specular is used in the sense of viewing (to have a wide view and to speculate is to look into, examine, observe etc.) then multi-specular or multi-viewing would seem to cover all aspects with the result that the phrase multi-specular radar (MSR) is used here and is intended to cover all aspects of radar imaging that can be combined to produce a new view whether it is frequency, polarization, look direction, or depression angle. The potential of MSR analysis was first noted by Hunting Technical Services Ltd during work connected with the European SAR-580 campaign.

The SAR-580 project in Europe described earlier provided the opportunity to obtain images in more than one radar frequency and more than one polarization. It also meant that some of the multi-frequency images were also flown in two sorties to obtain cross swath imagery over an area to give multi-look direction results. Over one particular area a number of flight directions were used so that as many as five different aspects could be compared.

The two main test sites for calibration purposes were Bedford Airport operated by the Royal Aircraft Establishment in the UK and Oberpfaffenhofen near Munich in Germany operated by DFVLR. At calibration sites, regular flight controlled imagery was flown of the sites using all the frequencies and polarizations.

Representative images of each waveband for each area were sent out as examples to the various researchers. The imagery was optical only at that stage, nevertheless the author was able to digitize the imagery through a television camera into a digital analysis system HIPAS where the imagery could be viewed on a screen. One of the benefits of an image analysis system is that images can be warped to fit each other, thus it was possible to fit three radar images of the same area taken at different frequencies together, assigning each to one of the colours on the television monitor to present a combined coloured image in much the same manner as is used with Landsat.

The result is shown in Figs 10.3 and 10.4 of Oberpfaffenhofen. This early

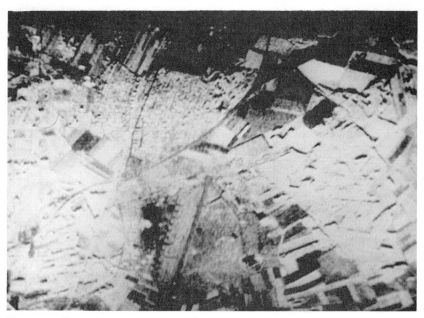

Fig. 10.3   Part of a single strip image (X-HH) used to construct a digital mosaic.

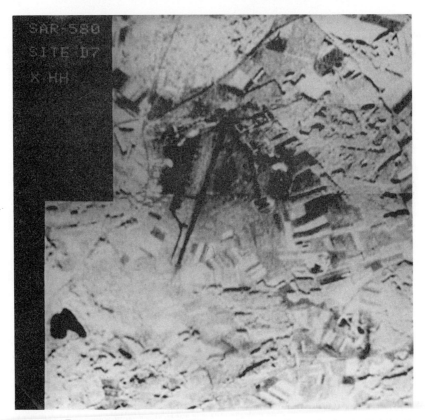

Fig. 10.4 Completed mosaic of X-HH, note how by using only the central portion of each strip, the radiometric dark areas at each edge due to antenna are eliminated.

attempt produced two conclusions; the main result was that fields showed as different colours and these colours could be found to relate to crop differences; and secondly that the effect of speckle was less intrusive than in the single image. This latter fact is not surprising if one considers the method of reproducing colours used by printers, a sharp colour image in a text book or quality magazine may look like a photograph but under a lens it will be seen to be constructed from a regular pattern of dots some being red, some blue and some yellow to give a coherent scene. Thus the speckle effect combines to produce a more coherent image when colour is used. The two factors in combination meant that the interpretability for the purposes of vegetation studies of the imagery was improved. Nevertheless for competent analysis of the frequencies it is desirable to remove the speckle.

Subsequently improved optical imagery of the Oberpfaffenhofen test site was obtained; this consisted of some six or more images as strips to cover the whole test area. The strips were all in the same flight direction and were available in X-, C- and L-band and in HH and VV polarization. In the first instance all the strips were digitized through the television camera into the HIPAS system, then the strips were fitted each to its neighbour in the same frequency and polarization to form a number of mosaics. These mosaics were then re-adjusted in order that they related to a base map and thence to each other. The result was six images which could be fitted to each other and observed in a number of colour combinations.

Plate 3 is one of the final mosaiced images. When comparing the combined HH polarizations with the combined VV image one area clearly showed fields on one image that were not apparent on the other. The fields on the HH are indistinct whereas on the VV they are clear and demonstrate distinctive field differences. The reason for this difference is not known with any certainty, there is some speculation that it is related to the crop type and row direction but this is not proven.

Other aspects related to this test site are described later in this chapter. In the use of MSR described only optical imagery was used, subsequently digital data became available and these were used to form combinations including combinations with other imagery, reference is also made to these later in this chapter.

The study described in the Section on Forestry for Thetford Norfolk UK (Churchill and Keech, 1983, 1984) describes how combinations can be made using fairly simple methods. In the event that the interpreter does not have access to computer equipment, the potential of such simple approaches should not be ignored. Using a simple back projection device with a combination of conventional 35 mm slide projection and colour filters will certainly improve interpretation as compared with studying each individual scene.

The major problem in producing a colour composited scene is the rectification of the image to a common control base such that each image will achieve an acceptable level of registration. The problem is less acute using space borne data than with airborne data especially if the compositing is between varieties of airborne data.

In analysing an image the most difficult problem then lies in removing the unwanted speckle in order to achieve a smoothed or more even tone image. The easier approach is to apply a filter to the image, in a basic form a median filter is usually employed and these may be 3×3, 5×5 or 7×7 windows. In the median filter the filter defines the representative value of a set of pixels as the median of that set, the set being a block 3×3 pixels and so on. If the window or pixel set is too large then most of the image detail is lost.

Using the image analyser it is possible to list various combinations of images colours and colour balance until combinations result which are meaningful to the interpreter. There can be no hard rule as to the best set of combinations to use for visual interpretation, much will depend upon the experience of the interpreter, their aesthetic preference and ability to discern colour. For an automated classification rules will have to be defined. From then onwards interpretation is a matter of correlating colour differences to ground and other data in order to arrive at a classification. The image can then be classified on a training area to produce a specific colour coded map. Output and hard copy by normal photowriters, electrostatic printers or ink jet plotters will provide a map type image; it is possible to produce reasonable thematic maps from combined radar data using, for example, an Applicon ink jet plotter (Kessler and Jano, 1985).

10.5.2   Field or other edge boundary effects

In interpreting cropped areas, smoothing often removes the boundary edges and in any case the boundary edge can have an effect on the internal pixel values, thus a hedge with large trees can give rise to a bright backscatter return to produce a boundary width disproportionate to the scale width of the boundary. If the areas being studies are in a developed country then it is probable that reliable maps already exist of the area showing the field boundaries. In interpreting or classifying crops then the principal interest is in the contents of a field and not the field boundary (although there is a case for recording changes in boundaries). Based upon this premise consideration should be given to introducing the field boundaries into the image. The most efficient way to achieve this is to digitize the field boundaries using a conventional flat bed or scanning digitized grid assigning an increased width to the boundary (this compensates eventually for any extended signal effect on the radar). An alternative method is to redraw the boundaries using a thick line and introduce the result into the computer via a video digitizer (Wooding, 1985; Mégier et al., 1985 and Churchill and Trevett, 1985). When combined the result is an image in which edges between fields have been replaced by a solid definite boundary leaving the internal image area to be analysed and classified (Plate 4).

Using this method it has been found possible to improve the internal classification reliability of crop units and at the same time, after colour coding, to produce an improved map image.

10.5.3   Other techniques

Between the visual interpretation of photo image and computer analysis there is another approach that can be used to give a more measured analysis. This may be desirable since it is known that the ability of an interpreter to identify and match colours or grey units is influenced by the surrounding tonal values. A unit surrounded by other strong colours or tones can take on some of the total value of those tones to result in a change of apparent tone or colour on the observed unit. Thus an interpreter may well identify two units as being comparable when precise densitometer readings could show that they are in fact different.

An instrument such as a Quantimet can be used to measure image density and derive values for each unit, using this system it is possible to identify major crops with an improved reliability over visual interpretation (Gombeer, 1985).

10.6   THE POTENTIAL FOR LAND USE

At the beginning of this chapter it was stated that the results and potential of SAR for vegetation surveys should be equated against the results from visible band imagery. The various experiences with SAR data tend to show that there is a good ability and potential to discriminate crops and vegetation units and that under certain conditions a high degree of reliability can be obtained.

Experience with MSS data whether the limited bands of Landsat MSS or the more extended range, up to 11 bands, of an airborne scanner does not demonstrate a substantial improvement over these accuracies. Indeed tests using the 11 channel scanner tend to show a reduction in accuracy for a wide range of classifications, although single unit classification may be more reliable. One reason appears to be that working with so many channels can introduce a form of confusion index when a whole range of differences can be obtained which even a skilled interpreter may have difficulty explaining.

In the use of combinations of SAR imagery it has been found that although the colours may change, as may the intensity of difference, the units distinguished remain very much the same, pointing to a greater consistency of classification, (Churchill and Trevett, 1985).

One particular study on the European SAR-580 project was able to compare multi-frequency SAR data with MSS data taken at the same time as the radar overflight and with detailed field results combined with reliable cadastral maps (Mégier et al., 1985). The cadastral maps were digitized to provide control points for rectification and definitive field boundaries. The images, airphotos, SAR and MSS, were all rectified to fit to this base.

The SAR data were processed to increase the number of looks to give two look data and these further processed to reduce speckle. For this the team devised an algorithm for adaptive filtering which was applied to the data. In this approach the filtering technique smooths homogeneous areas while at the same time preserving contrast at the edges thereby enabling augmentation techniques to be used in the same manner as for optical imagery. The SAR was available as X-band as HH and HV and C-band as HH and HV.

After smoothing the SAR data a wide variety of combinations were tried, combinations of SAR, of MSS, and of SAR with MSS and air photos. The first tests showed that the air photographs showed the lowest discrimination potential and they were subsequently eliminated as an interpretation aid.

The results of the study are quoted here since they are the results of one of the only valid comparative studies made.

(1) In almost all cases the results are good or very good for three class classification discrimination (maize, grass, cereals).

(2) The visible bands MSS 8 and 10 are not better, rather they are worse, than SAR bands XHH and CHH both for three class and five class discrimination.

(3) Bands XHV and CHV used together are not so good but nevertheless contain discrimination potential. When added to bands XHH and CHH, isolated or together, they improve the discrimination performance, especially for five classes, the best results being obtained with four SAR bands.

(4) The addition of bands MSS 8 and 10, isolated or together, to SAR bands does not always improve results, it can even cause them to deteriorate. The results are however improved by adding the two bands for five class discrimination.

(5) The discrimination in five classes is rather good by using four SAR bands and very good using four SAR and two MSS bands.

(6) The best results are obtained by using four SAR and two MSS bands, with average discrimination performances better than 90% for three or five classes. SAR data can also be combined with infra-red data as well as MSS, the addition of infra-red may improve some aspects of classification although rectification remains a problem. The microwave data improve classification compared to visible/infra-red models and the differences observed are also related to biomass and to crop height. In both SAR and infra-red the data are influenced by soil moisture with SAR also being sensitive to surface texture which can be significant in a tilled field and cause confusion with a cropped area.

There is therefore evidence to show that SAR is at least equal to MSS for the discrimination of vegetation and has the added advantages of an all weather capability.

The full potential of SAR has yet to be realized, whether through acquisition instrumentation, processing or analysis, the potential of land use studies could prove to be better than the potential of optical systems and there is growing evidence to this effect.

REFERENCES

Attema, E. P. W. and Ulaby, F. T. (1978) Vegetation modelled as a water cloud. *Radio Science*, **13**, (2).

Bush, T. F. and Ulaby, F. T. (1976) Radar return from a continuous vegetation canopy. *IEEE Trans.* **AP-24**, (3).

Bush, T. F. and Ulaby, F. T. (1977) *Cropland inventories using satellite altitude imaging radar.* RSG Technical Report, University of Kansas Centre for Research, Lawrence, Kansas.

Bush, T. F. and Ulaby, F. T. (1978) An evaluation of radar as a crop classifier. *Remote Sensing of the Environment* Vol. 7, Elsevier, pp. 15–36.

Churchill, P. N. and Trevett, J. W. (1985) *An evaluation of SAR-580 multi-frequency radar data over the Norfolk test site.* SAR-580 Final Report.

Crown, P. (1980) *Evaluation of radar imagery for crop identification.* Intera Environmental Consultants' Report ASP-80-1, Ottawa.

DeLoor, G. P. and Hoogeboom, P. (1982) Dutch ROVE program, *IEEE Trans.*, **GE-20**, (1), 3–10.

Endlicher, Gossman, Mauser and Parlow (1985) SAR-580 Investigators' Final Report.

Engheta, N. and Elachi, C. (1981) *Radar backscattering from a smooth surface with a vegetation layer cover.* IGARSS 1981 Proceedings.

Fung, A. K. (1979) Scattering from a vegetation layer. *IEEE Trans.*, **GE-17**, (1).

Fung, A. K. and Fung, H. S. (1977) Application of first order renormalization method to scattering from a vegetation like half space. *IEEE Trans.*, **GE-15**, (4).

Gombeer, R. (1983) *Crop and land-use classification study on SAR data over Belgium.* European SAR-580 Investigators' Interim Report (ESA).

Gombeer, R. (1985) *Crop and land-use classification study on SAR data over Belgium.* European SAR-580 Investigators' Final Report (ESA).

Hoekman, D. H., Krul, L. and Attema, E. P. W. (1982) *A multilayer model for backscattering from vegetation,* IGARS 82 Proceedings, IEEE.

Hoogeboom, P. (1983) Classification of agricultural crops in radar images. *IEEE Trans.,* GE-21.

Kessler, T. and Jane, A. (1985) *Results of an evaluation of digital SAR-580 data for land use identification.* SAR-58- Investigators' Final Report.

King, G. I. (1980) *Radar discrimination of crops.* Intera Env. Consultants Ltd, Report ASP-81-1, Ottawa, Canada.

Lang, R. H. (1981) Electromagnetic backscattering from a sparser distribution of lossy dielectric scatterers. *Radio Science,* **16**, (12).

Lang, R. H., Saker, S. S. and LeVine, D. M. (1982) *Scattering from a random layer of leaves in the physical optics limit.* IGARSS, TA1-3.

Lang, R. H. and Sidhu, J. S. (1983) Electromagnetic backscattering from a layer of vegetation: a discrete approach. *IEEE Trans.,* **GE-21**, (1).

Le Toan, T. (1982) *Active microwave signatures of soil and crops: significant results from three years experiment.* IGARSS TP-2, June 1–4.

Mégier, J., Mehl, W. and Ruppelt, R. (1985) *Methodological studies of rural land use classification of SAAR and multi-sensor imagery.* European SAR-580 Investigators' Final Report.

Mehta, N. (1983) *Crop identification with airborne scatterometers.* IGARSS, PS-2, San Francisco.

Paris, J. F. (1982) *Radar remote sensing of crops.* IGARSS, FA4.

Paris, J. F. (1983) Radar backscattering properties of corn and soybeans of frequencies of 1.6, 4.75 and 13.3 GHz, *IEEE Trans.*, **GE-21**, (3).

Pei-Yu, J., Deli, G., Guixiong, S., Junnyu, M. (1983) *Radar backscattering coefficients of paddy fields.* IGARSS, FR-5.

Protz, R. and Brisco, B. (1980) *Analysis of soil and vegetation characteristics influencing the backscatter coefficient of a SAR system.* Intera Env. Consultants Ltd, Report ASP-80-1, Ottawa, Canada.

Sieber, A. J. and Trevett, J. W. (1983) Comparison of multifrequency band radars for crop classification. *IEEE Trans,* **GE 21,** (3).

Shanmugan, K. S., Ulaby, F. T., Narayanan, V. and Dobson, C. (1983) Identification of corn fields using multidate radar data. *Rem. Sens. Env.,* **13,** (3).

Ulaby, F. T. (1981) Microwave response of vegetation. *Adv. Space Research,* 1, 50–70.

Ulaby, F. T., Aslam, A. and Dobson, M. C. (1982) Effects of vegetation cover on the radar sensitivity to soil moisture. *IEEE Trans,* **GE 20,** (4).

Ulaby, F. T. and Moore, R. K. (1973) *Radar sensing of soil moisture.* Proc. 1973 International IEEE-GAP and USNC/URSI meeting.

Ulaby, F. T., Moore, R. K. and Fung, A. K. (1982) *Microwave Remote Sensing – Active and Passive,* Addison-Wesley Inc., New York.

Waite, W., MacDonald, H., Tolman, D., Barlow, C. and Bonengasser, M. (1978) *Dual polarised long wavelength radar for discrimination of agricultural land use.* Proc. ASP, Albuquerque.

Wooding, M. G. (1985) *SAR-580 radar for the discrimination of crop and crop conditions.* SAR-580 Investigators' Final Report.

11

Satellite systems

11.1 INTRODUCTION

The term 'satellite system' is being used here to include systems operated from the orbiting shuttle missions as well as independent rocket launched orbiting satellites. A number of lunar missions and missions to other planets have used microwave technology in order to observe the surface of the planet involved, these are not included in this evaluation although the interpretation, principally of the structure and geology, uses techniques and experience gained from earth observation.

At the present time there has only been one orbiting radar satellite specifically designed for earth resources observations, this is the Seasat satellite. There have been a number of satellites launched for military purposes and included in these have been satellites launched by the USSR and known to be imaging microwave instruments deriving their power from nuclear power packs. No data are available for these systems and no results of interpretation for earth resources have appeared in the scientific literature.

11.2 SEASAT

The Landsat optical imaging satellites have very limited applications to the study of the world's oceans. When conditions have been suitable, i.e. no cloud cover, images have been obtained of the near polar regions which have been of value in observing sea ice and iceberg formation from glaciers; in shallow coastal zones the optical imagery has been successfully used to provide data on depths by the analysis of colour changes. In the deeper areas the imagery is of little value and it certainly is of no value in the study of wave patterns.

It was against this background that the oceanographic community pressed for a satellite more suited to their particular requirements. It had already been demonstrated that data from a scatterometer taken over a sea surface, could provide information on wind speed and direction and that precise height measurements could be made using a radar altimeter. These instruments therefore provide the main pack of a potential satellite. Experiments from cliff edges and sea towers as well as airborne data had indicated that an imaging microwave instrument could record sea surface patterns including wave patterns of surface waves and internal waves, it therefore was desirable that the satellite should include an imaging microwave capability.

Figure 11.1 is an example of Seasat over a coastal area, the surface wave patterns, river outfalls, current flow and eddy patterns as well as an apparent relationship to sub-surface bathymetry can all be clearly seen. A Landsat scene of the same area would display no such distinctive features in the sea areas. This figure also demonstrates a typical standard image product for Seasat published by the ESA-Earthnet facility at Frascati in Italy.

The overall objectives of the satellite system are shown in Table 11.1 (taken from Satellite Microwave Remote Sensing edited by T. D. Allan) it will be seen that the final instrumentation included a visible and infra-red radiometer (VIRR), a scanning multi-frequency microwave radiometer (SMMR) as well as the instruments previously identified as desirable. The satellite was constructed and subsequently launched by NASA on 27th June 1978 and known as Seasat, the SAR configuration of Seasat is given in Table 11.2. In this instance the instrumentation is seen related to shuttle radars which are referred to later in this chapter.

Satellites are placed into a circular orbit around the earth, usually a near polar orbit, which means that they orbit at a slightly inclined angle to the equator (for Seasat this is 108°). As they circle in orbit the earth will rotate naturally within the orbiting circle producing consecutive passes of the satellite. In orbit the satellite will pass down one side of the globe and up the other side known as the descending and ascending modes.

Landsat, being an optical system dependent upon the sun for illumination, can only image when passing over the side of the earth which is in sunlight and this is so arranged that it occurs in the descending mode, it cannot therefore image in the ascending or unilluminated mode. A radar imaging device suffers no such constraints, it can acquire imagery in both the ascending and descending modes; furthermore, the inclination of the orbiting paths means that when passing the same area in the ascending and

Fig. 11.1   Seasat 1 synthetic aperture radar (SAR) image of the Isle of Wight. Data were acquired by Royal Aircraft Establishment (RAE) on September 2nd, 1978 at the Oakhanger Station, UK and digitally processed by Deutsche Forschungs – und Versuchsanstalt fuer Luft – und Raumfahrt (DFVLR) at Oberpfaffenhofen, Germany.

descending modes the orbit flight paths will intersect at an angle, thereby providing a dual directional look over an area.

We have already seen the effect of radar shadows and this dual directional looking provides the opportunity to overcome some loss of detail in radar shadow zones, it also provides the opportunity to examine target response from two different viewpoints.

Table 11.1   SEASAT objectives

ALT	To measure very precisely (~ 10 cm) the satellite altitude above the sea surface.
	To measure the significant wave height of the ocean surface at the subsatellite point.
	To utilize the altitude measurement, confirmed with precision OD, to extract oceanographic and marine geoid information.
SAR	To obtain radar imagery of ocean wave patterns in deep oceans.
	To obtain ocean wave patterns and water-land interaction data in coastal regions.
	To obtain radar imagery of sea- and fresh-water ice and snow cover.
SASS	To deduce local wind vector information from ocean radar scattering coefficient information.
	To obtain synoptic ocean radar scattering coefficient measurements over a wide variety of sea and weather conditions and instrument parameters.
	To obtain radar scattering-wind vector interaction data over the ocean.
SMMR	To measure ocean surface temperatures.
	To measure ocean surface wind speeds.
	To measure liquid and vapour water in the atmosphere.
	To measure ice coverage and ice characteristics.
	To provide propagation corrections for other experiments.
VIRR	To provide image feature identification (land, clouds, etc.) in support of other experiments.
	To obtain thermal images of the ocean for various oceanographic purposes.

ALT   Radar altimeter
SAR   Synthetic aperture radar
SASS   Wind scatterometer
SMMR   Scanning multi-channel microwave radiometer
VIRR   Visible and infra-red radiometer

As with any orbiting satellite the orbit cannot be restricted to either land or sea, but must overfly both, thus although Seasat was designed as an oceanographic experimental satellite it could nevertheless be used to obtain radar images of the land mass for resources studies.

Unfortunately Seasat had an unexpectedly short life and failed completely in October 1978 after 1503 revolutions, the failure is thought to have been due to a total loss of power.

The short life span meant that many of the stations planned to receive data

Table 11.2 Summary comparison of principal characteristics of SEASAT and SIR L-band (23 cm) imaging radars

	SEASAT	SIR-A	SIR-B
Active period of satellite	June–October 1978	November 1981	August 1984
Orbit inclination	Polar 108°	38°	57°
Coverage	c. 100 km² × 10⁶	c. 10 km² × 10⁶	c. 30 km² × 10⁶
Altitude	794 km	c. 260 km	225 km
Radar frequency	1.275 GHz	1.282 GHz	1.282 GHz
Polarization	HH	HH	HH
Radar pulse bandwidth	19 MHz	6 MHz	12 MHz
Antenna dimensions	10.74 × 2.16 m	9.35 × 2.16 m	10.7 × 2.16 m
Swath width	100 km	50–55 km	Variable 20–55 km
Peak power transmitted	1 kW	1 kW	1 kW
Depression angle	70° ± 3°	43° ± 3°	Variable 15°–60°
Azimuth resolution	25 m	40 m (6 looks)	25 m (4 looks)
Range resolution	25 m	40 m (slant range)	Variable 58 m (at 15°) – 17 m (at 60° incidence) (ground range)
Number of looks	4	4–7	4
Processing	Optical and digital	Optical only (8 hours)	Optical and digital (8 hours + 25 hours)
Digital link capacity	? 640 k bits/s	N/A	46 m bits/s
Attributable cost	$400 million	$9 million	? $13 million (exclusive of launch costs)

had not become operational and many experiments had to be abandoned. The USA receiving stations (Alaska, Florida and California) did operate, as did the Canadian station in Newfoundland and the only station outside the North Americas to receive data was the Royal Aircraft Establishment, Oakhanger, receiving station in England. The latter recorded 272 minutes of data from 53 passes which included data across Iceland and Europe, although in the last case not complete cover.

Various experimenters were able to use the imagery in order to evaluate its potential for a number of resources applications. Even six years after the failure of the satellite, the data are still being used for experimental research and have proved invaluable. Papers on findings still appear in the literature and it is now recognized that the satellite was extremely successful in obtaining good data despite its short life span.

Satellites need to telemeter data back to earth which means that all data originate in digital form, thus for Seasat there was a large amount of digital data and the possibility of repeated coverage either along the same pass or as a result of ascending and descending mode image to give a cross swath view. Seasat therefore provided a rare research opportunity.

The Seasat imagery acquired over Europe is shown in Fig. 11.2. This represented some of the most significant data obtained and the results of the various experiments conducted on these data throughout Europe, including the digital processing, are comprehensively described in 'Satellite microwave remote sensing' edited by Allan (1983) which is the assemblage of the papers presented by the Seasat Users Research Group of Europe (SURGE) at a specially convened conference.

The coverage for N America is shown in Fig. 11.3 (taken from the Manual of Remote Sensing) where it will be seen that an even greater amount of cross swath data were received.

The Seasat data are archived by NOAA in the USA and Earthnet in Europe and are available to researchers and others through the normal channels. The existence of this data base is significant both for its historical data content and for comparison with subsequent systems. Data are available either optically processed or digitally processed. This book is primarily concerned with land resources and it is the results obtained with the Seasat SAR overland which are of particular significance in this context, the flight diagrams show that a considerable amount of overland images were obtained.

11.2.1 Geological studies

A number of studies were undertaken using Seasat imagery for geological studies, one of the most comprehensive studies was the study completed by Hunting Geology and Geophysics Ltd for the RAE and NASA. These

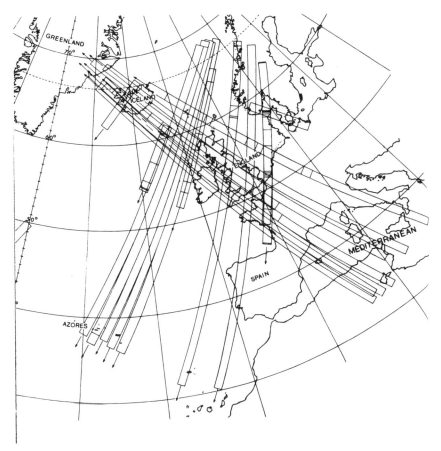

Fig. 11.2 Seasat coverage of Europe obtained over a 4–5 hour total operational period. (From Allan, 1983.)

studies encompassed Iceland and the United Kingdom. Almost complete cover was obtained over Iceland and similarly for the UK. Figure 11.4 is the Seasat SAR mosaic constructed for the UK which, given the scale it can be reproduced at in this book, can still demonstrate the potential of a satellite SAR system for imagery, the mosaic clear shows both sea data and land data, note also how the industrial midlands complex around Birmingham and Manchester shows up clearly, in fact is more identifiable than Greater London. The inclusion of Iceland in the study provided a completely different geological area, one which has not, for example, been subjected to extensive surface modification by man.

In the United Kingdom large areas have been modified by glaciation and ice action with the result that the expression of the bedrock lithology was

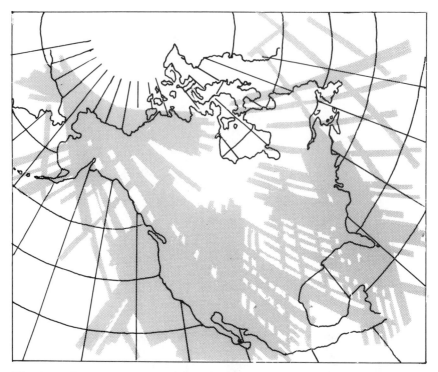

Fig. 11.3   Seasat coverage of N. America from 29th June 1978 until 10th October 1978 when the satellite failed. (After the Manual of Remote Sensing.)

masked. In other areas substantial cultural development also masked the underlying geology. As a result the Seasat radar was found to be of little value in such circumstances in aiding lithological distinction. It was useful in displaying lineaments and in comparison with the existing geological maps the radar interpretation appeared to show significantly more lineaments.

A comparison of the interpretation with the existing geological maps can be seen in Figs 11.5 and 11.6, the geometry of the radar is not in precise relationship to the map projection and a best overall fit has been used. In Fig. 11.5 a major joint system has been identified and in Fig. 11.6 metamorphic structures are clearly evident.

Where a geological unit is interpreted on the basis of a distinctive drainage pattern or a clear unit such as sand dunes, interpretation can be well correlated with the existing map. In other areas where the surface has been extensively modified by man the exact lithological boundary cannot be determined but units can be separated on the basis of tone and texture which relate to the known geology. In one area 25 units could be discriminated on

Fig. 11.4 Seasat image of the United Kingdom.

tone, texture and shape and these are shown in Table 11.3 as an example of how determinative criteria are used to separate units for classification. Table 11.4 is the known geological classification of the area shown related to the units in Table 11.3.

Once again the radar demonstrated its potential to present linears despite the modified land surface, the linears on the radar were much more numerous than were depicted on the corresponding geological map. The area in the west of England has been surveyed geophysically and therefore the radar interpretation could be checked against the known geology and the geophysical map data, demonstrating how all available knowledge has to be taken into account in interpretation. This latter comparison showed that there was a correlation between linear patterns and electromagnetic data. It was possible to use the digital data on a precision interactive image analyser and Fig. 11.7, an area in the south west of England, shows the result after it has been stretched to improve contrast and after it has been two-look processed. The area shown is west of Plymouth and covers an area of granite moorland rising to a maximum of 400 m. Only one geological unit could be defined, the Bodmin Moor Granite, however the linears are particularly evident and in the lower right corner there is one of many circular features identified on the imagery.

Through the use of the image analyser it was possible to compare Seasat with Landsat, Fig. 11.8 is such a comparison over Plymouth town photographed from a split screen display on the analyser. The comparison showed that the railway bridge over the River Tavy can be clearly seen on the radar but not on the Landsat, also the Seasat has picked up ships at anchor but the vessels are 'mothballed' or permanently moored in the river thus there is no question of changes due to time. Cultural features are better defined on the Landsat and the vegetation, which is well depicted on the false colour image, could help identify geological units, however very little structural detail can be determined compared with the Seasat.

The Iceland imagery offered the advantage that it presented an area with exposed geology which had been well mapped. The location near to the polar regions meant that there was a higher probability of cross swath imagery and large overlaps between passes which could produce stereograms. The study showed that the majority of the major units mapped by the Icelandic Geological survey could be recognized by their geological expression and Fig. 11.9 is the radar geological map that was produced out of the study.

An example of the interpreted imagery is given in Fig. 11.10. This image also demonstrates some of the effects of high relief with the bright line of the layover area of the ridge edges. Unit 7 is Rhyolite, a Pleistocene volcanic unit which forms a sharply incised unit of moderate topography with fewer lineaments than the adjacent Palasonite formation (3). Unit 9 are Holocene

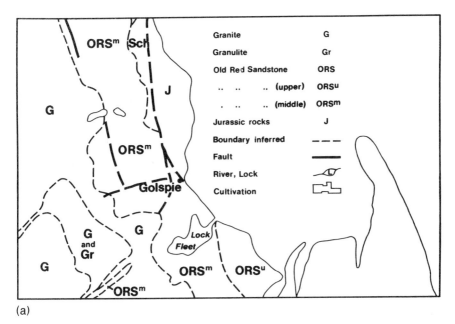

(a)

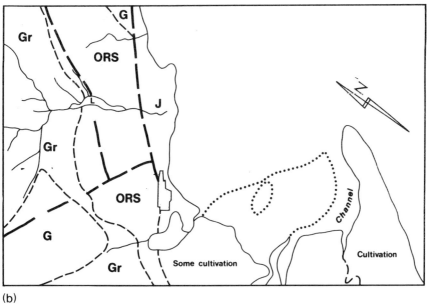

(b)

Fig. 11.5 A comparison of geological interpretation (1). (a) The geological survey map. Geological survey of Great Britain sheet 5 (1″ to 4 miles). (b) Interpretation of Seasat. Geological interpretation: Dornoch Firth Area. (Hunting Geology and Geophysics Ltd.)

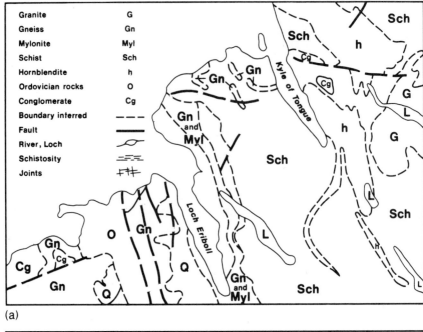

(a)

Legend:

Granite	G
Gneiss	Gn
Mylonite	Myl
Schist	Sch
Hornblendite	h
Ordovician rocks	O
Conglomerate	Cg
Boundary inferred	- - -
Fault	▬▬▬
River, Loch	⌇
Schistosity	═╤═
Joints	⊢╫

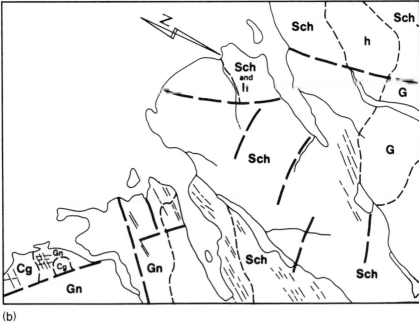

(b)

Fig. 11.6 A comparison of geological interpretation (2). (a) Geological survey of Great Britain, Sheet 5 (1″ to 4 miles). (b) Geological interpretation: Cape Wrath and Whitten Head. (Hunting Geology and Geophysics Ltd.)

Fig. 11.7   Seasat imagery of SW England after digital enhancement to improve contrast.

basalt lavas, the light toned uniform fine grain textured units with no fine lineaments. One of the many prominent dykes has been identified on the figure.

Figure 11.11 is part of the icecap and Fig. 11.12 is an enlargement of the SW corner where a volcanic cone and crater can be clearly distinguished with the form of the cone making an excellent demonstration of the effect of radar illumination. The bright side faces the satellite and produces a high signal return with a decrease in backscatter around the cone to a partial shadow side.

A comparison of Landsat and Seasat could be made: Fig. 11.13 is the Landsat scene and Fig. 11.14 is the same area on Seasat. The Landsat shows more features at the foot of the glacier in the top right and better drainage features, while old moraine lines can be seen on the Seasat image.

The geological study indicated that the Seasat imagery radar can be successfully interpreted for geological structures and morphology and in particular the expression of linears often reveals data that are not discernible on any other imagery. Where the land surface has been modified then lithological identification is not reliable. For geological studies interactive analysis of the single image did not produce any significant benefits. The

Table 11.3 Southwest England radar units: their determinative criteria and expression of geology

| | | | Determinative criteria | | Lineaments | | | |
	Drainage incision	Relief	Distinctive morphology	Vegetation patterns	Major	Minor	Expression of geology	Stratigraphy
Unit								
A	Present	Moderate	Dissected plateau		Present		Moderate	L–M Devonian
B	Present	Modest	Subdued ridges	Present	Few	Abundant	Imprecise	L–M Devonian
C	Absent	Modest	Irregular hummocky				Good	Coastal aeolian sands
D	Slight	Negligible	Undulating undissected				Good	Coastal aeolian sands
E	Slight	Low	Open rolling	Common	Few		Poor	Carboniferous culm measures
F	Slight	Modest	Open rolling	Common	Numerous		None	No consistent geologic correlative
G	Slight	Modest	Boldly hummocky	Common	Numerous		None	Carboniferous culm measures
H	Slight	Modest	Dissected rounded	Present	Present		Poor	Carboniferous culm measures
J	Strong	Considerable	Bold rounded landforms	Few	Abundant		Moderate	Granite
K	Strong	Considerable	Bold rounded landforms	Few	Abundant		Moderate	Granite

L	Slight	Low	Hill slopes	Very few	Very few	Poor	Dolerite, culm measures
M	Present	Modest	Dissected rounded	Present	Few	Poor	Culm measures
N	Slight	Modest	Little feature	Abundant	Present	Poor	Permian marls, sandstones, breccias
O	Present	Modest	Irregularly hilly	Abundant	Present	Poor	Granite, carboniferous, oligocene
P	Present	Modest	Local ridge	Common	Present	Poor	Permian, Tertiary
Q	Absent	Lacking	Flat			Good	Quaternary
R	Absent	Negligible	Subdued irregular			Good	Aeolian sands
S	Slight	Moderate	Hummocky	Few	Present	Poor	Various
T	Slight	Low	Indiscernable	Few		Poor	L. Devonian
U	Present	Moderate	Ridge and valley	Few	Numerous	Imprecise	L. Devonian
V	Present	Moderate	Irregularly incised	Few	Few	Imprecise	L. Palaeozoic metamorphics
W	Present	Modest	Subdued	Present	Few	None	U. Devonian
X	Present	Moderate	Rounded	Few	Few	Poor	U. Devonian
Y	Present	Varied	Boldly rounded	Few	Few	Poor	Various
Z	Strong	Considerable	Boldly rounded	Present	Abundant	Moderate	Granite

Table 11.4 Southwest England: General stratigraphy, lithology and correlation with radar units

Stratigraphic units	Lithology and correlation with radar units	Radar unit
Recent/Pleistocene	Blown sand, alluvium, sandspits, bars etc.	C, D, R, Q
Tertiary	Recognizable from low or absent relief and context. Limited extent. A small unit (Haldon Gravel), identified by coincidence of plantation. Otherwise not recognizable.	–
Cretaceous	Not represented.	–
Jurassic	Not represented.	–
Permo–Trias	Marls, breccias, conglomerates. Partial undefinitive correlation with three radar units.	P, N, S
Carboniferous	Culm measures. Shales, slates, sandstones and grits. Partial undefinitive correlation with several radar units.	E, H, M, O, Y
Devonian – upper	Slates, lavas, sandstones, shales, limestone. Undefinitive correlation of part with one radar unit	X
middle, lower	Slates, grits, conglomerates, shales, limestones. Some correlation with geology in one unit (A). Other units reflect imprecise and poor correlations.	A, B, S, T, U
L. Palaeozoic metamorphics	Schistose rocks. Some correlation but imprecise with one radar unit.	V
Intrusive rocks	Granite batholiths of Bodmin and Dartmoor. Three distinctive radar units giving moderately accurate delineation of occurrences.	J, K, Z
Uncorrelated	Some areas on the map have not been assigned to any of the radar units described. Some radar units (F, G and W) are not geologically correlated.	Unassigned and F, G, W

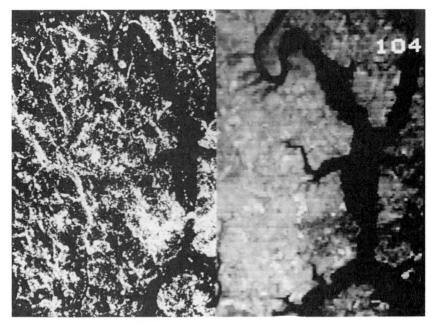

Fig. 11.8   A comparison of Seasat and Landsat.

Seasat radar did not perform as well as Landsat but the radar was configured for oceanography not land studies and for geology the need for regular repeat imagery is not so acute.

In studying the significance of linears in interpretation caution has to be exercised in their directional balance, the bias of radar has already been referred to and Fig. 2.22a is a typical rose diagram of the frequency and orientation of linears on a typical Seasat geological scene, Fig. 2.22b by comparison is a rose diagram from a Landsat image. It will be seen that Landsat with its effect of sun angle shows a similar bias but in a different segment.

A number of similar geological studies were carried out in N America mainly be research scientists at the Jet Propulsion Laboratory (JPL) at Pasadena. The data were used to produce uncontrolled mosaics of California, Pennsylvania, Florida and Jamaica.

A comparative study of Seasat SAR images and Landsat images over the Appalachians showed that small scale linear topographic features were preferentially enhanced on the radar images (Ford, 1980, Seasat bulletin 1980). In this instance observation of cross swath data over the area helped to eliminate the linear bias introduced by scene illumination and the final complication of linears was made using both coverages. This clearly demonstrates the potential value of an orbiting system where data in both

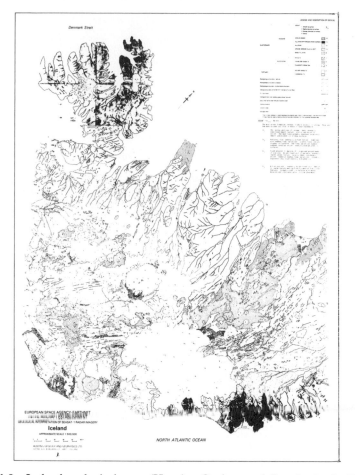

Fig. 11.9   Iceland geological map. (Hunting Geology and Geophysics Ltd.)

the ascending mode and descending mode can be used to provide the interpreter with imagery from two different view points.

In another study in the USA it was found that the high return from the numerous steep drainage channels together with the backscatter from the vegetation and rocks served to observe the subtle geological features that were being sought.

Even the oceanographic aspect of Seasat can be of value to the geologist. It has been shown how the radar imagery in coastal areas reflected the subsurface topography, this can be of value in areas where offshore exploration activities are planned. In a similar vein the study of the sea surface state under a variety of conditions, the paths of currents and features which can influence both the currents and the sea state are important factors

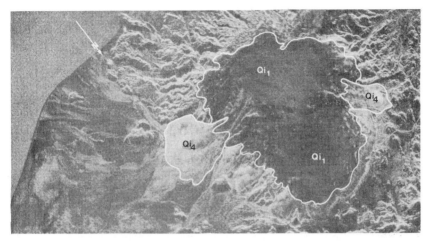

Fig. 11.10   Part of interpreted image of Iceland. Qi_1 smooth ice surface, Qi_4 glacier.

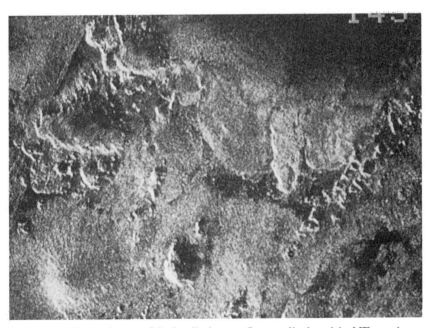

Fig. 11.11   Seasat image of Icelandic icecap. Icecap displayed in NE quadrant. Volcanic cone and basalt lava field visible in SW quadrant. Pronounced linear structures in NW quadrant reflect the Mid-Atlantic Ridge alignment.

Fig. 11.12 Volcanic cone on Seasat. Enlarged extract showing part of volcanic cone and crater. The diameter of the crater is probably in the region of 200 m. The image is oriented with highlit face uppermost in order to obtain the effect of topographic relief when viewing. There is some diminution of information in the highlighted face.

in deciding on the location and type of drilling rig and subsequently of value in planning pipelines and oil installations. The sea surface dynamics also have to be studied for they would play an important role in the distribution and effect of any possible oil pollution resulting from damage to rig or pipeline.

11.2.2 Land use applications

The Seasat study referred to earlier undertaken by Hunting Geology and Geophysics included a study of potential land use applications which was undertaken by a sister company, Hunting Technical Services Ltd. The study was undertaken for areas in England and other countries in Europe, the modified land cover referred to by the geologists became an advantage as there was a mixed agricultural and land use pattern to study. Digital image

Fig. 11.13 Landsat image.

Fig. 11.14 Seasat image of same area as Fig. 11.13. Old moraine lines are discernible although drainage channels are imperfectly registered. (Area approximately 30 × 40 km.)

analysis was used extensively for the land use studies with a reasonable amount of success.

One of the main test areas selected for study was in the east of England in north Norfolk. The area was selected because the field sizes were above average for England. There was a limited range of crops and the area is relatively flat so that relief did not affect the imagery.

In Fig. 11.15 the raw Seasat data is shown for this test area with the appropriate histogram of the grey levels, in Fig. 11.16 the image has been contrast stretched and a geometrical correction applied. The benefits of digital analysis will be readily apparent, few field patterns are visible in the raw image and very little tonal discrimination of units, in the top right most of the detail is obscured by a large dark area; in the improved image, field and road patterns have become distinguishable and tonal differences between fields or units can be interpreted, in the previous dark area detail is revealed including the significant emergence of an air field. The strong north–south linear to the left of centre of the improved image could not at first be reconciled with the obvious expected details, in other words there was no main railway, canal or motorway in that area, in fact it is the line of the old Roman road. The road, Pednors Way, is no longer in use and its route is now marked by hedges, footpaths and farm tracks, however throughout its length the old embankments persist.

In Chapter 4 reference was made to multi-look processing and Fig. 11.17 is four scenes combined for comparison to show the relative effects of one look, two look and three look processing and three look processing with a low pass filter. Three looks certainly improved the sharpness and clarity of the image and was better for interpretation but four or more looks yielded no further benefit and could be considered detrimental. The use of a low pass filter further reduced speckle and improved the definition of the boundaries. As a result of this section of the study all the interpretation was subsequently carried out on a three look image with a low pass filter.

One of the objectives was to determine whether the radar could be classified using digital image analysis on a computer analyser and the next series of Figures 11.18 and 11.19 demonstrates how a key area was selected, enlarged, classified using the computer to analyse tone into colours related to crop or vegetation groups and the key analysis applied to the whole image. In this case the available imagery had to be used for the tests rather than obtaining imagery at a specific time that could be planned to coincide with optimum crop conditions and field studies. It had been planned to carry out field studies at an optimum time during a subsequent overpass but the short satellite life meant this opportunity was missed and it was a year before the area of imagery became known and the only field work possible was based upon farm records.

Comparison with known data showed a high degree of correlation with

Fig. 11.15 Raw Seasat digital image. Incorrect geometrical shape resulting from 4 × 2 sampling of the pixels. No stretches applied. Grey level histogram superimposed on the scene.

Fig. 11.16 Same scene as Fig. 11.15 after linear stretch. Linear stretch: correct geometrical shape with 4 × 3 sampling. Linear stretch applied to expand the levels between 9 and 114 over the range 0–255. Grey level histogram superimposed on the scene.

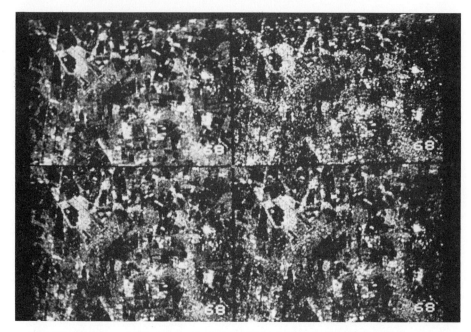

Fig. 11.17 Comparison of single and multi-look processing. Comparison of one look (top right), two look (bottom right), three look (bottom left) and three look imagery with low pass filter applied (top left).

the crops and vegetation (about 60%) but there could be no further breakdown of units using computer techniques so that on balance it was considered that image analysis by computer gave no advantage over optical interpretation. It has to be remembered that this was a single image, studies described for airborne studies in an earlier chapter (Chapter 10) showed that a combination of look angles and/or frequencies or polarization could yield more discriminations of units. On this basis future satellite data can be expected to improve on these findings.

An area near Bristol in the west of the United Kingdom which included the Severn River provided the opportunity for a multi-temporal comparison, the area was one imaged in the crossing of the ascending and descending modes. The resulting images are given in Figs 11.20 and 11.21. The example provides an excellent demonstration of the type of deduction that can lead to wrong conclusions. The Severn River is well known for its great tidal variation and for the sudden surges of flow that can result from the effects of exceptional tides. The first assumption was that in Fig. 11.20 the river was full but with little or no wind to disturb the water and the tide coming in the flow was quiet whereas in Fig. 11.21 the river was low and the bright returns were coming from the rough surface of the sandbanks. A subsequent study

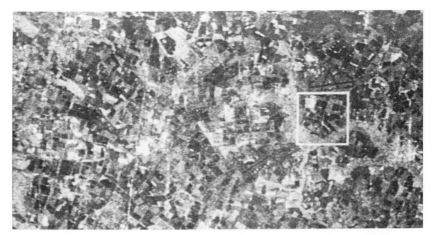

Fig. 11.18 Full image with area selected for detailed study. All these are photographs taken of the viewing screen of the image analysis system and not a subsequent hard copy output. In this view the scene to be analysed is shown, and superimposed on the image during the analysis is a location box which can be moved over the image to a suitable location and its size varied in each direction as required.

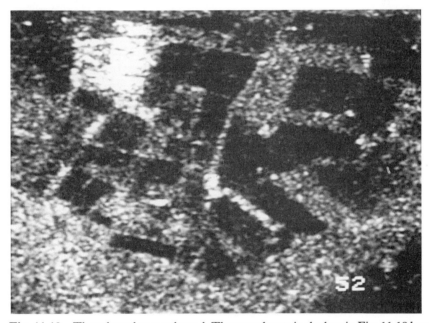

Fig. 11.19 The selected area enlarged. The area shown in the box in Fig. 11.18 has now been extracted and enlarged to fill the whole screen and uses all the available digital data that are going to be analysed using conventional density slicing techniques, none of the other techniques used on conventional Landsat MSS analysis can be effectively used on a single band image.

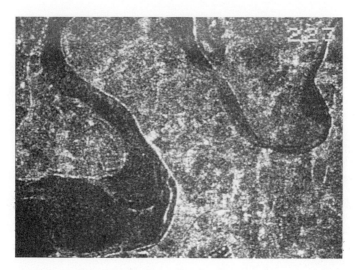

Fig. 11.20 Seasat of river estuary (1). The black area is the Severn River, it was interpreted as the river at low tide when this section of the river has exposed mud banks. It was assumed that both the mud banks and the water would give a low backscatter and thus appear uniformly black. In reality the river was at high tide with the tide backing up the river so that water was from bank to bank. It so happens that the river flow was reduced at the end of a summer dry spell and there was little or no wind to disturb the water surface.

Fig. 11.21 Seasat of river estuary (2). The same area as in Fig. 11.20, originally it was thought that this was a low tide scene with the bright areas due to sand banks. Subsequent study of the tide tables showed that the river was at full tide and the areas of high backscatter return were commensurate with a strong river flow giving rise to disturbed water on the bends.

of the tide tables for the time of overflight revealed that the situation was reversed, Fig. 11.20 showed the river at low tide when the mud flats and the quieter water would give the same low backscatter while in Fig. 11.21 with the tide running out the high backscatter was coming from the swift river as it curved round the bend. This is a clear demonstration of how without adequate local knowledge or field observations a completely erroneous interpretation can be made.

The multi-temporal data afforded the opportunity to examine whether changes in crop state could be identified. Two images of different acquisition dates were superimposed in the computer and were displayed on the monitor one in red and one in green, where no change has occurred the image is a mix of the two colours, where change does occur either red or green predominates. Three fields were identified where change had occurred, field studies determined that between the imaging dates the wheat in fields in these areas had been harvested so that there was a difference between cropped land and bare soil. There is some indication therefore that radar imagery could be of value for observing changes in land use.

Comparisons between Landsat and Seasat were made for land use purposes and an example is shown in Fig. 11.22 where the improved resolution of radar over Landsat can clearly be seen and Seasat is shown to

Fig. 11.22   Seasat compared with Landsat of the same area.

be a much better quality imaging system than is at first apparent. The use of multispectral data in a colour composite improves the appearance of the Landsat and obviously more recent images using a thematic mapper will show a marked improvement.

The studies of European areas gave very similar overall results but only in one area was any field work possible. By a fortunate occurrence some field work, albeit of a historical nature, was possible on a study area near Turin in Italy. This latter was valuable in explaining two apparent anomalies observed on the imagery.

In the first instance it was difficult to reconcile the interpretation of wooded areas with the areas marked as woodland on the local detailed map. The signature for woodland could clearly be established from comparison with known woodland on the existing land use maps but this left a number of other larger areas of apparent woodland which were not marked as such; field work not only confirmed the existing woodland interpretation but also revealed the additional woodland to be poplars which are grown as a crop in the area and therefore not considered as permanent woodland. In another area the bright returns were indicative of buildings, probably industrial, but the maps showed a rural area, field work revealed a new industrial estate under construction. Other excessively bright returns from field areas were identified as being maize at maturity. The results are quoted here as demonstrations of the essential importance of adequate supplementary knowledge and ground checking of interpretation before the final pronouncement on classification is made, and of how major land use changes from rural to industrial can be reliably identified.

The final example of Seasat imagery used in this study is given in Fig. 11.23. The discussion on the Iceland image indicated that the northern site had a better chance of producing overmapping orbit lines, this proved to be the case and two strips are given in the figure which can be viewed stereoscopically. Some of the layover effect can be seen and although the three dimensional view is of interest it adds little to interpretation for geology or land use and is of insufficient reliability for topographic mapping.

In evaluating Seasat it has to be recognized that the SAR was optimized for oceanographic studies and not intended for studies overland. While it provided invaluable experience in the study of the earth's surface from radar imagery it cannot be considered as a definitive verdict on the potential of such systems for land applications, only a system optimized for land studies will provide these results.

Other researchers have examined Seasat for various land use applications, some work was undertaken in the USA on urban studies and agriculture but with mixed results. Of greater interest was the hydrological implications of imagery obtained over Iowa in the centre of the USA. A major storm had moved across the area the day prior to image acquisition and storms of

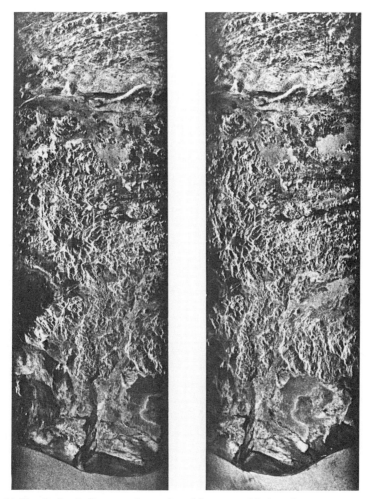

Fig. 11.23 Iceland. Stereo pair produced by same side imaging.
Left: revolution 590, sub-swath 3.
Right: revolution 633, sub-swath 1.

various types continued up to 10–12 hours prior to image acquisition. Just before satellite overpass the storms broke up and indicated rain cells moved across the area towards the north west. The image clearly has a light area and a dark area, the light area is in the area of principal rainfall and storm activity whereas the darker areas are where rainfall was absent, and these can be related to the path of the subsequent rain cells breaking off and moving across the area.

The ability to identify the effects of such localized precipitation activity and to possibly monitor the paths of rainfall is an important advance in hydrology, enabling patterns to be better studied and data referenced to crop yield prediction programmes.

11.3   SHUTTLE IMAGING RADAR – SIR-A

Although Seasat subsequently failed and no direct successor was launched there was, however, a back-up instrument built. It was decided that this radar should be flown in one of the Shuttle Space Orbiter programmes, this time not so much as an oceanographic experiment but as an experiment in assessing radar for geological mapping. SIR-A was fitted into the space shuttle Columbia and launched in November 1981 as part of the Office of Space and Terrestrial Application first payload known as OSTA-1.

Like Seasat the SIR-A radar is a synthetic aperture radar operating in L-band at 1278 MHz corresponding to a 23 cm wavelength about a million times that of visible light. The seven panel antenna has a total length of 9.5 m and was fixed relative to the shuttle at an angle of 47° to the left of the sub-nadir track corresponding to an incidence angle of 50° at the centre of the swath. This angle was selected in anticipation of eliminating some of the effects of the low incidence angle (23.6°) at the swath centre of Seasat, it should be remembered however that this angle had been selected for oceanographic and not geological studies.

The SIR-A data were optically recorded onto film carried in a cassette. Figure 11.24 shows the coverage obtained by the shuttle during its flight; unfortunately due to technical problems, the shuttle flight was curtailed, however this did not seriously affect the SIR-A programme. Imaging of ten million square kilometres was acquired between 41° north and 36° south and covered both land and ocean. It will be seen that the swaths covered large areas of the more unsurveyed areas of the world and included a great number of wet tropic areas where the acquisition of Landsat imagery had been hampered by the existence of almost perpetual cloud cover.

The signal film was processed at the Jet Propulsion Laboratory (JPL) at Pasadena USA. It was optically correlated onto 13 cm film. The imagery had a swath width of 50 km and a resolution of the order of 40 m. (Figure 11.25a and b demonstrates the shuttle technique.) In Seasat the low incidence angle referred to earlier meant that the backscatter was dominated by slope and topography, the new higher incidence angle reduced this effect and the backscatter was more sensitive to surface texture thus improving interpretation.

To obtain a nominal ground resolution in the order of 25 m a SLAR radar at a wavelength similar to SIR-A would require an aperture of 2 km, rather more than shuttle could carry, hence SAR is the only viable system for space

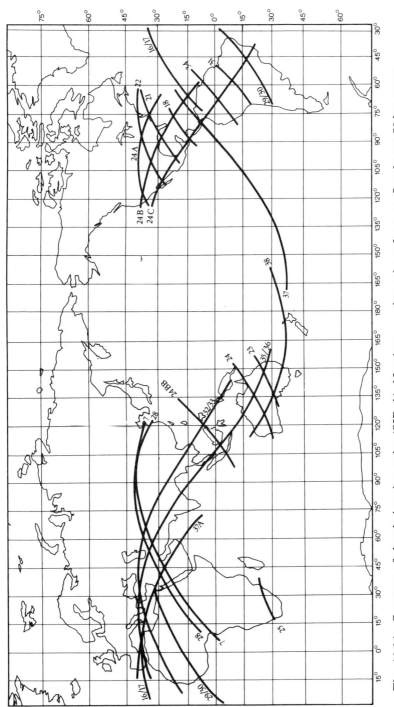

Fig. 11.24 Coverage of shuttle imaging radar (SIR-A). Numbers are data take references. Swaths are 50 km across.

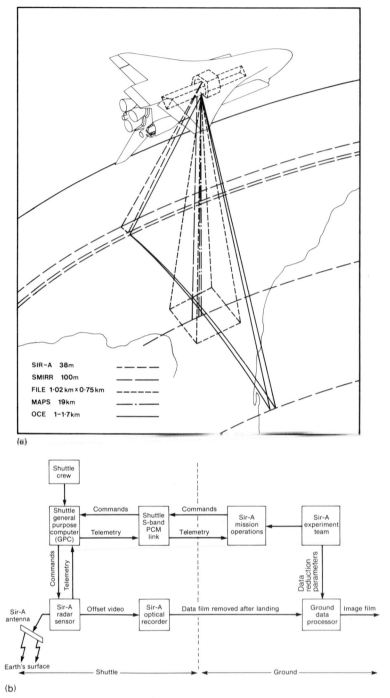

SIR-A   38m   — — —
SMIRR   100m   — — —
FILE 1·02 km x 0·75 km   — — — —
MAPS   19km   — · — ·
OCE   1–1·7km   ————

(a)

Fig. 11.25 The space shuttle SIR-A technique and coverage. (a) The shuttle operation. (b) Functional diagram for the SIR-A system. (Climino and Elachi, 1983.)

imaging. Using SAR the resolution was obtained using a 10 m antenna to give a resolution three times that of Landsat.

Some of the figures quoted by Elachi of JPL demonstrate the problems of digital image processing. He estimates some 1000 complex operations were required to synthesize the long aperture and to determine the brightness of a single pixel of 25 m. Given that the satellite was moving at 7.5 km s^{-1} and that a swath width of 100 km on the ground was observed, approximately a billion operations were needed to generate a high resolution image from one second of data collection. To this has to be added the need to calculate Doppler shifts, altitude distortions and space craft altitude effects. Elachi states that one of the fastest computers currently available, the CRAY-1 can do about 100 million operations a second so that any real time processor would need to be an order of magnitude faster to cope. The fastest computer available to process satellite radar has a throughput of one to 500, i.e. 1 s of data collecting will require 500 s of processing. JPL is aiming to develop a processor that will keep pace with the processing requirements. It is however easy to see why, at present, optical processing is used.

The backscatter return is influenced by the roughness of a surface and the angle of the beam to that surface. In the latter case changes in the angle of the surface will result in the same effect, so that small changes in angularity or slope of an otherwise homogeneous surface will produce a change in signal. If this change forms a pattern and is not random then the subsequent pattern may be related to a subtle surface expression that might otherwise go undetected by optical imaging methods. It is for this reason that in heavily vegetated but otherwise homogeneous covered areas such as the large Amazonian forest regions, radar can be used to detect the underlying geomorphology.

Similarly changes in roughness in surfaces can also be detected, changes which do not record with the same strength, if at all, on optical imagery. Where the two combine the expression is even more pronounced, thus small slopes may change the soil moisture content and thereby the overlying vegetation growth to result in both a change of angle and of subtle roughness to produce a strong indication of the presence of the slope.

Unlike optical imagery, radar microwaves can penetrate certain materials and the depth of penetration is proportional to the wavelength of the signal. Radar cannot however penetrate water, which optical imagery can, and it is the variance in penetrability in an otherwise uniform matrix which is used to detect variance in soil moisture content. In very arid areas, i.e. deserts, there is no soil moisture and as a result there is some sub-surface penetration. Figures 11.26 and 11.27 are two images of the same area, the upper from the metric camera and the lower from SIR-A. SIR-A clearly records the presence of sub-surface drainage features. Their presence can be detected on the photograph, but the expression on the radar is stronger and more

Fig. 11.26   Metric camera scene.

Fig. 11.27   The scene in Fig. 11.26 on SIR-A. (European Space Agency.)

immediately obvious. Some of the drainage features, long suspected by geologists, are as wide as the present Nile valley. A similar difference can be observed when comparing with a Landsat image.

The expression of the drainage is probably a sum of relevant factors, for a start the radar has an improved resolution over Landsat MSS, (although this is not true for the metric camera) then the drainage represents changes in slope, subtle though they may be, and depositional differences would give rise to small changes in grain size or roughness and finally there is penetration. What proportion of each is responsible for the effect it is impossible to determine, certainly the summed result is convincing. Researchers looking at images of Borneo in the wet tropics have been able to delineate swamp areas despite the vegetation cover, giving rise to the theory that some penetration of the vegetation occurred. Subsequent field studies carried out in Egypt showed that penetration of at least 1 m was achieved in some areas and that in some cases this could increase to 2 m.

A special study of the potential of SIR-A imagery for geological mapping was made by a number of researchers at the International Training Centre (ITC) in Enschede, Holland, who used imagery from various areas in the world: Argentina, Brazil, Thailand, China, Turkey and Greece included.

In general the overall research results emphasized the value of radar in providing an improved expression of topographic relief and of linears and the researchers were of the opinion that the SIR-A imagery was of value in regional mapping, especially if that imagery could be allied to Landsat imagery (Koopmans et al., 1983).

A feature of all of the conclusions was the emphasis on the use of other data sources to supplement the interpretation, these ranged from Landsat to aerial photographs and to existing geological maps and field data. The latter, the use of existing geological maps, provided a means of evaluating the radar interpretation and where this was possible a comparison of linears showed compatibility of trends with more linears being identified from the SAR. The existence of forest cover in some of the test areas did not detract from the value of SAR, especially in lineament studies.

The Seasat SAR depression angle was 70° but the SIR-A, although using basically the same radar, had a depression angle more suited to geology at 43° with the result that the layover effect was less intrusive except in areas of exceptional relief. A comparison of SAR with aerial photographs showed that the SAR gave a much improved expression of relief than the air photo and a comparison of SIR-A radar with airborne and band radar (Pedreira, 1983) showed that the feature shadows on the SIR-A radar gave a better indication of the dip of beds, this was in an area in Brazil where airborne imagery was available. The overall result of the indication of the dip of beds was that folded structures were more easily detected whereas on the photography they were virtually undetectable. A comparison table (Table 11.5) drawn up from the study compares various forms of imagery for interpretation and this is reproduced here as it provides an interesting comparison.

In the study of a volcanic area in Turkey a comparison of SAR with Landsat was made, the results showed that the radar gave a better expression of lineaments, however the comment was made that 'Landsat and radar imagery should be considered and used as complementary and not competing techniques. If used together, or even superimposed, the results of analysis will be enhanced'.

Figure 11.28 is a dramatic example of SIR-A imagery obtained over the western Sinkiang area in China, this was one of the areas used in the ITC studies and Fig. 11.29 is the geological map for the same region. The study

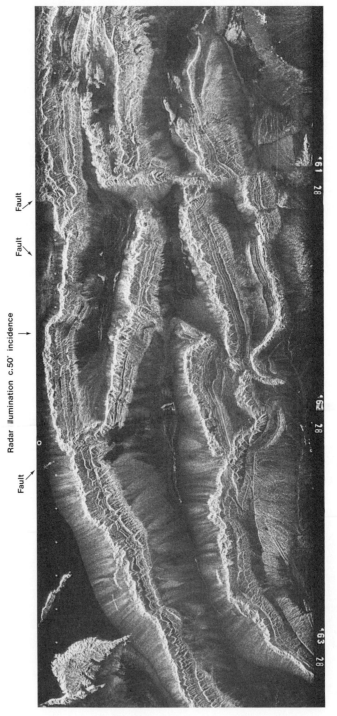

Fig. 11.28 SIR-A image of Sinkiang area, China.

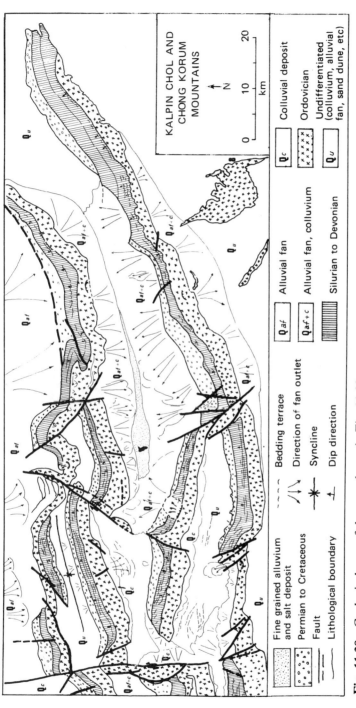

Fig. 11.29 Geological map of the area shown in Fig. 11.28.

Table 11.5   A comparison of radars for geology

Feature	SAR (SIR-A)	Airborne X-band SAR	Landsat MSS-5	Landsat MSS-7	Air photographs
Synoptic cover	G	F	G	G	P
Relief	G	G	P	F	G
Tone	F	F	G	P	G
Lithology	G	G	P	F	G
Folds in crystalline rocks	F	P	P	F	G
Faults	G	G	P	G	G

G Good; F Fair; P Poor (after Koopmans et al., 1983).

again compared Landsat and SIR-A for interpretation and concluded that SIR-A showed excellent lithological and structural characteristics – superior to those obtained from Landsat, that the well bedded gently dipping slopes were better detected on SIR-A but were not apparent on Landsat, while the unconsolidated Quaternary sediments could be mapped on both systems.

Perhaps one of the most interesting comparisons was the comparison of lineament interpretation from SIR-A with Landsat and the geological map for the area. A study undertaken at the Jet Propulsion Laboratories (JPL) compared SIR-A with Landsat and with the existing geological map. It was immediately evident that both Landsat and SAR identified a much greater number of lineaments, with the SAR giving the most, and the peak direction of the geological map interpretation could be identified. The difference in directional bias and intensity reflected the directional bias of the two systems used in an orbiting satellite; where the possibility of two look directions could be obtained much of the directional bias of the SAR could be compensated for.

Various studies of SIR-A imagery, mainly for geological studies, have been made throughout the USA and by US researchers on areas elsewhere in the world. The most significant of these will be the work carried out in the desert of north Africa following the identification of antique drainage channels. Field studies have been undertaken to test the existence of the channels and to compare the grain size of the overlying and sub-surface sands.

A study over the Appalachians, referred to earlier in the section on Seasat, showed that the SIR-A was able to depict more short linear features than the Seasat which is attributed to the look angle difference, they also noticed the enhancement of small scale topography on the SIR-A as opposed to the Seasat and this endorses the Dutch findings described earlier.

Figures 11.30 and 11.32 are further striking examples of SIR-A imagery. Figure 11.27, which is part of the Kelpin Tagh uplift region in NW China, shows how large scale faulting is clearly visible and that the shales, limestones and stratified sandstones on the ridges can be clearly identified. An example of a vegetated area is given in Figs 11.30 and 11.31 which is part of the Pakaraima Mountains of western Guyana where, despite the cover, the structures can clearly be seen, especially the dissected plateaus.

The image of the area around Chiran in China (Fig. 11.32) is one of the most striking demonstrations of the population density of that country. Each of the bright spots is a village and also field patterns are clearly visible. The signal resembles the high return of villages observed in Nigeria where the signal was a result of corrugated iron being used for roofing, however, it is not known if this material has been used throughout these Chinese villages. The villages clearly have a characteristic which makes them excellent radar targets, even so the density of habitation is extraordinary. Compare this image with Fig. 11.33 which is the capital of Brazil, Brazilia; the town is clearly visible but does not demonstrate the intense returns of the Chinese villages. Figure 11.34 compares airborne SLAR with shuttle imagery showing the difference in resolution.

Experiments have been conducted on the potential of satellite SARs for mapping (Curlander, 1984) using Seasat and SIR-A imagery. The research concentrated on the construction of digital mosaics using using adjacent strips and identifying common control or tie points along the strip overlap to control rectification. As a step in the mosaic construction the images had to have the distortions derived from azimuth skew and ground range nonlinearity, a system was devised whereby a specially written algorithm was written and applied to the digital image data prior to mosaicing. The secondary distortions resulting from relief displacement would require adjustment to a digital terrain model and this has yet to be achieved. This line of research is one of the topics being addressed by the SIR-B mission.

11.4   SIR-B

In 1984 a further shuttle mission was flown known as SIR-B. The objectives were to continue geological studies and by varying the depression angle to test the effects of changes in depression angle on the geological expression (Tables 11.6 and 11.7 and Figs 11.35 and 11.36).

A full programme of research was planned for SIR-B which included global sites, sites in Europe and sites across N America. Unfortunately the mission was beset with problems, for example it had been hoped that the mission would be the only imaging system on this particular shuttle, but, for

Fig. 11.30 Western Guyana/eastern Venezuela: Rio Mazaruni, Pakaraima and Merume Mountains, Proterozoic Clastic Roraima formation forms massif slabs with steep scarps, basaltic lava flows and sills at A and partially waterlogged savanna (dark areas) of La Gran Sabana at B. Scale approximately 1:500 000.

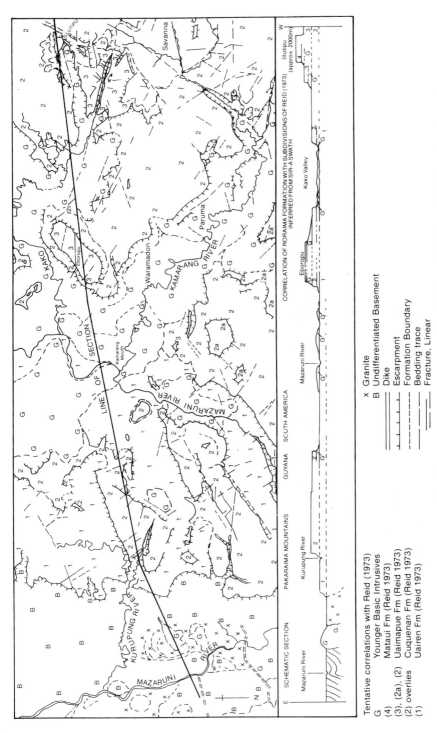

Fig. 11.31 The traverse of the northern Pakaraima Mountains, Guyana by SIR-A subdivision of the Roraima formation.

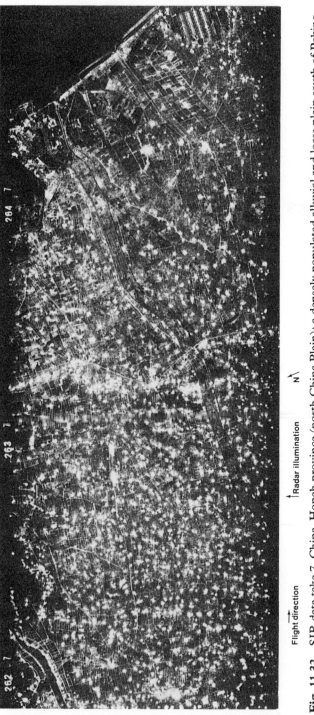

Flight direction

↑ Radar illumination

↑ N

Fig. 11.32 SIR data take 7. China, Hopeh province (north China Plain): a densely populated alluvial and loess plain south of Peking. The bright radar reflectors are towns and villages, probably with metal roofs. Roads show clearly as moderately bright. The dense network of field boundaries shows probably because the roads and tracks are raised slightly above the flat waterlogged fields showing black. Note that the fields are rectangular evenly across the whole image; there is no obvious near range compression or distortion because the angle subtended by the swath edges at the spacecraft is small (2°–3°). The parallelogram shape of the fields near the coast (left of image) is probably genuine. The scale is 1:500 000 at centre of swath but varies by about 5% between the swath edges being smallest nearest to the radar look direction. The apparent resolution of the image is approximately 40 m × 40 m but the roads (below 10 m in width) with their linear form and distinct contrasting radar characteristics can be readily resolved. Individual radar reflectors in the towns are not resolved.

Fig. 11.33. SIR-A image of Brasilia, Brazil.

a number of reasons, it eventually had to share the mission with the large format camera LFC, a photographic experiment. Placing the LFC on the same mission as SIR-B meant imaging time had to be reduced and split between the two experiments.

Furthermore it was decided that it was impossible to carry adequate tapes to record all the data and that direct data transmission would be used via a relay satellite back to the USA. During flight the first problems occurred when the radar antenna which had been folded in order to allow two satellites for launching to be accommodated, would not open and had to be assisted by the astronauts. Then it was found that the directional antenna used to point to the relay satellite was loose and would not stay oriented. This now meant that only the tape recorder could be used on a much reduced imaging programme. Subsequently it was found that the whole space craft could be used to orient the relay antenna, so that each tape loading of data could be relayed to earth. The overall result was a much reduced imagery programme and the cancellation of many experiments. It was subsequently found that the antenna had a loose nut and it is an example of how vital attention to detail can be at every level.

At the time of writing the research on the data from SIR-B has just commenced but there is every indication that this will yield as many exciting results as previous campaigns. In some of the first tests a special method of colour enhancement based upon density slicing has been used which has

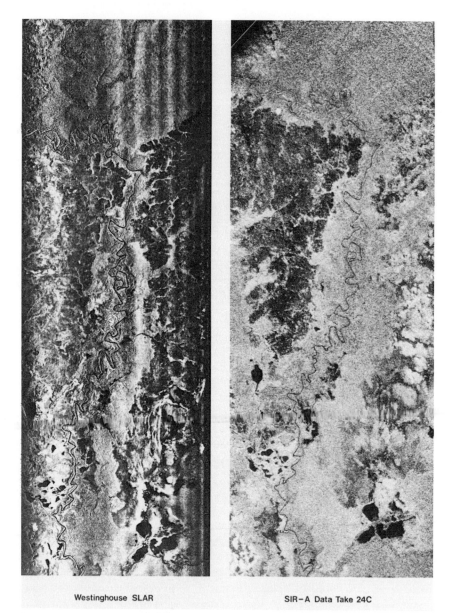

Westinghouse SLAR   SIR–A Data Take 24C

Fig. 11.34   A comparison between airborne SLAR and SIR-A.

Table 11.6 Standard product specification – SIR-B

Parameter* Product	Ground range resolution (m)	Azimuth resolution and number of looks	Radiometric calibration	Image geometry	Image location	Image media	Dynamic range Point targets (dB)	Dynamic range Distributed targets (dB)	Data quantity and processing period
Quick-look digital	100	100 m 1 look	None	Slant range with Doppler	None	Print on TV monitor	30	10	Several frames during mission
Survey optical	30–100	50 m 4 looks	None	Slant range with Doppler	None	Positive film and print, 5 in	17	12	All 25 h (digital) during first 6 months
Flight optical	30–100	40 m 6 looks	None	Slant range within 38 of 500 000 : 1 map	None	Positive film and print, 5 in	17	12	8 h during first 12 months
Standard digital	15–55	25 m 4 looks	1 dB (relative)	Correct to within 100 m, 10 m pixels sampled cross track and along track	To 1 km†	100 km frames on 1600 bpi and 8 × 10 film and print	45	10–30 dependent on look angle and number of bits	Minutes; first 6 months 4 h : 1985 6 h : 1986

* Variations are functions of look angle.
† Location and geometric errors are primarily due to shuttle orbit knowledge errors.

(JPL Publications 82–78, 1982)

Table 11.7 SIR-A and SIR-B sensor characteristics

Parameters	SIR-A	SIR-B
Orbital altitude (km)	260	225
Orbital inclination (degrees)	38	57
Frequency	1.28	1.28
Polarization	HH	HH
Look angles (degrees)	47	15.60
Swath width (km)	50	20.50
Peak power (kW)	1	1
Antenna dimensions (m)	9.4×2.16	10.7–2.16
Antenna gain (dB)	33.6	33.0
Bandwidth (MHz)	6	12
Azimuth resolution (m)	40(6 look)	25(4 look)
Range resolution	40	58–17
Optical data collection (h)	8	8
Digital data collection (h)	0	25
Digital link capability (mbits s^{-1})	N/A	46

(JPL Publications 82–78, 1982)

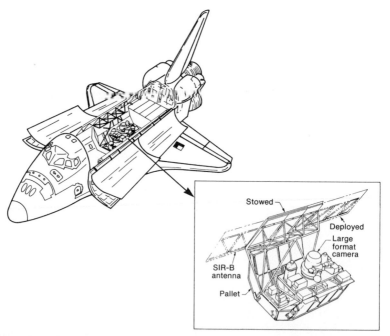

Fig. 11.35 The SIR-B instrumentation shown on the shuttle pallet together with the large format camera, a parallel mission to obtain stereoscopic optical images on film, the configuration is shown in the orbiter and in detail in the inset.

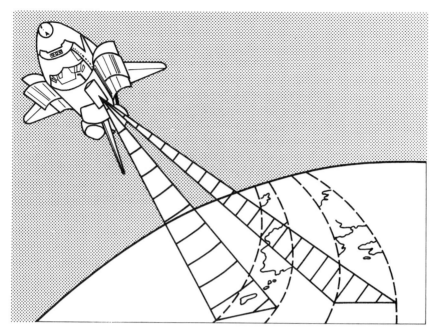

Fig. 11.36   The SIR-B concept of variable depression angle.

been found to add to the geological interpretation potential by pointing to possible lithological units.

These three major space SAR programmes have been highly successful in demonstrating the potential for SAR imagery over land, mainly in the field of geology, but the land use applications look equally encouraging.

The only other space SAR experiment that has been conducted was a German built system which was flown on a shuttle mission in 1984 known as the MRSE (microwave remote sensing experiment), unfortunately there was a power failure and the system failed to operate. The instrument has since been returned to Germany and it has been found that the fault lay in one of the wave tubes. Subsequent testing has replicated the fault and it is hoped to improve the instrument design and reliability so that it can be flown on another shuttle mission, possibly in 1985.

REFERENCES

Allan, T. D. (1983) (Ed.) Satellite Microwave Remote Sensing, Ellis Horwood.
Ford, J. P. (1980) Seasat orbital radar imagery for geologic mapping: Tennessee-Kentucky-Virginia. *Amer. Assoc. Petrol. Geol. Bull.*
JPL Publication 82–78 (1982).

Koopmans, B. N., Pacheco, Woldai and Payas (1983) *Sarthi, a side looking radar survey over the Iberian pyrite belt*. The European SAR-580 Investigators' Preliminary Report. September 1st, The Joint Research Centre ESA.

Pedreira, A. J. (1983) Spaceborne SAR imagery interpretation Babia Area, Brazil, *ITC Journal*.

Seasat orbital radar imagery for geological mapping AAPG bulletin V 64 (1980) N12, pp. 2064–94.

Manual of Remote Sensing (1983) (Ed. R. K. Moore), 2nd edn, American Society of Photogrammetry.

12

Future trends and research

12.1 INTRODUCTION

The development of new techniques in aerial photography may not have ceased but the operational development has stabilized and in other areas of optical imaging new developments are being concentrated on scanning devices and linear arrays. Conventional optical interpretation has also stabilized and the principles of computerized image analysis are basically well understood although researchers continue to try to achieve automated classification. In microwave technology, although airborne imaging techniques have been successfully used commercially for some 15 years, in comparison the science is only beginning and many of the techniques are only just being understood.

The understanding of the potential of SAR imagery is becoming more widespread and the pace of research is quickening with a consequent growth in the development of new techniques. The announcement of SIR-A opportunities for researchers attracted a handful of researchers, by the time, three years later, that SIR-B was announced, NASA was overwhelmed with applications to conduct research projects and consequently they had to carry out a severe selection process, such is the pace of activity.

This chapter may be unusual in this type of book but it is necessary to look at some of the directions research is taking, for a start it may help to assure resources scientists that the use of SAR has a future and that its potential can only increase, it may also encourage those same interpreters to enter upon more detailed research projects.

Much of what is being written here will be out of date in a very short time although the lead time before the launch of a new satellite tends to be almost ten years and once the system specifications have been determined there is very little room for change if the launch deadlines are to be met.

12.2   AIRBORNE SARS

The most recent airborne system is the IRIS introduced by MacDonald Dettwiler Associates, and the upgraded SAR-580 from the Earth Resources Institute of Michigan. There are no major system developments envisaged for some years, although system improvements are obviously possible. As with any products, future development will be dependent upon a suitable market response, systems are being developed for military use but their specifications and operation will remain classified for some time. Interpreters can only take into consideration available technology and until the use of SAR for practical resources surveys expands the state of the art will probably remain static for some time. It is possible that the advent of satellite SARs will give an impetus to the use of airborne systems, this is a trend that has resulted from the use of Landsat and strangely seems to be related to the improvement in resolution of the satellite data.

In the case of Landsat, it was eventually recognized that some studies required imagery more frequently than could be provided by satellites or in more wavebands or at a larger scale or finally were too small for special imaging requests to be considered for satellite operation. As a result airborne multi-spectral scanners have begun to find a role of their own.

There is evidence that interest in airborne SARs is increasing, the unreliability of space systems to date may be a major influence in this changing attitude. The power requirements of SARs are a constraint for space work and these constraints are less of a problem for airborne systems, especially if multi-frequency systems are shown to be of particular benefit.

The major problems related to improving airborne systems may be found to be more in aircraft operation than the radar itself. For example it has been shown earlier in this book that many of the system errors in airborne data acquisition result from aircraft altitude variations, from the inability to maintain and record a known attitude above the land surface being imaged and from positional errors in navigation. Optical imaging systems which use scanning or linear array devices can exhibit many comparable distortions attributable to the same flying variations. It is not unreasonable to assume that there will be improvements, if not in the stability of aircraft platforms, at least the accurate recording of flight variations will be sought. The recording of height data can be assisted by airborne radar altimeters, however these have to be calibrated against the flying altitude of the aircraft in order that the variations in flying height and variation in terrain can be separated. In many areas of operation there are no reliable maps and certainly a larger number of areas for which no accurate or reliable surface height data exist.

It is possible that the advent of stereoscopic images from optical space systems such as SPOT will enable improved topographic maps to be constructed supplying a basic terrain model for height determination. Satellite position fixing systems already exist which can ensure an aircraft's position is accurately recorded and these systems will improve in accuracy and capability enabling the airborne surveys to be reliably positioned. The improvement of flight data recording will enable the performance of existing SAR systems to be optimized, only then will it be feasible to look for improvements in the systems themselves.

Both the updated SAR-580 and the IRIS systems now include a recording referencing system whereby flight data and radar gain settings are recorded at pre-determined intervals and these intervals are also recorded with the SAR data. By this means it is possible to inject into the radar data the relevant signal data and flight parameters so that the image processing can take these into account.

The SAR-580 series of campaigns led to the realization that the use of multi-frequency radars could improve interpretability. The problems that arose derived from the failure of the systems themselves, not all frequencies produced the same performance and levels of image quality. Already (1985) the system has been updated and substantially improved so that the C-band is now capable of imagery comparable in quality to the X-band. Reference has been made to the building of an improved C-band system and it is known from the instrument designers that research is being concentrated towards improving the packaging of systems such that they occupy less space, make a more efficient use of available power and operate with greater reliability.

The fact that improvements in airborne systems can and are being made, should not be taken as evidence that airborne SAR is not yet of use. The case studies quoted in this book are evidence of their effective application and the new systems are capable of producing imagery which can find immediate application to a wide range of disciplines in a number of circumstances.

12.3 SPACEBORNE RADARS

Another shuttle experiment is planned known as Samex for Shuttle Active Microwave Experiment. This has as its mission objectives to provide multi-parameter SAR imagery for the scientific community, to determine the information content in SAR imagery and to develop new SAR technology applicable to free flying or satellite SARs; Samex will be used for vegetation discrimination, geological mapping, hydrological studies and ocean wave spectra studies. There will be two flights each with (Table 12.1) 50 hours of data collection.

Table 12.1   SAMEX proposed instrument profile

Launches in 1985 and 1986	
Orbital inclination	50° polar
Orbital altitude	200 km
L-band (HH) and C-band (HH and HV)	
Incidence angles	20° to 70°
Resolution	10–30 m
Swath width	35–125 km

The next major satellite operation will be the European radar satellite
ERS-1 due for launch in 1989. Figures 12.1–12.4 show the proposed
configuration of this satellite, its data network, payloads and coverage
footprint. It is intended that ERS-1 will have a two year life cycle and that
the ERS series will be a continuing series in order to provide a reliable and

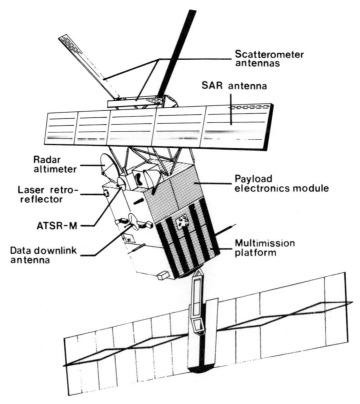

Fig. 12.1   The ERS configuration.

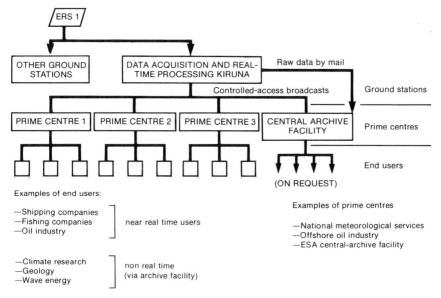

Fig. 12.2   The ERS-1 data distribution concept.

constant satellite radar facility. ERS-1 is primarily configured as a sea observation satellite and will carry a radar altimeter and scatterometers. The imaging radar will have as its specifications the parameters in Table 12.2.

Table 12.2   ERS-1 Proposed instrument profile

Spatial resolution	100 m × 100 m or 30 m × 30 m
Swath	80 km minimum
Radiometric resolution	1 dB for 100×100 m at $0^{-\circ} = -18$ dB
	2.5 dB for 30×30 m at $0^{-\circ} = 18$ dB
Incidence angle (on the ground) $\cong 23°$ at mid swath	
Frequency 5.3 GHz in HH polarization	
Mean RF power	< 400 W
Data rate	$\cong 100$ Mb s^{-1}
Antenna size	$\cong 10$ m × 1 m

The SAR will operate in C-band over the land and it is expected that this will be of major interest to many resource scientists. The ERS-1 will be launched from Kourou Space Centre (French Guiana) by the Ariane 2 (or 3) launcher and will use the same multi-mission platform as the Spot imaging satellite being built by the French. Spot is an advanced optical imaging satellite which will observe in three spectral bands with a ground resolution

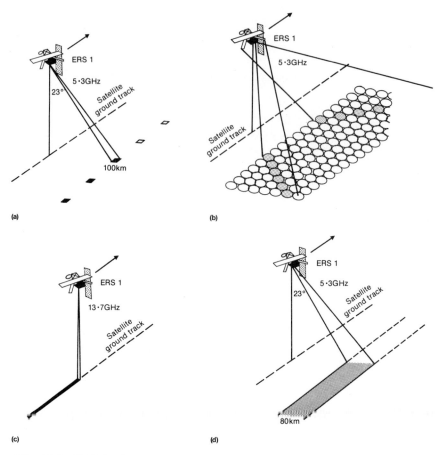

Fig. 12.3 ERS-1: the payload instruments.

(a) Wave scatterometer. The SAR will be used on a sampling basis as a wave scatterometer obtaining data for the determination of directional wave image spectra.

(b) Wind scatterometer. Designed to obtain all-weather day and night measurements of surface wind speed and direction.

(c) Radar altimeter. Measuring the distance between the satellite and the ocean surface and extracts the significant wave height.

(d) SAR mode.

of 20 m or in panchromatic black and white at 10 m. It is the first optical satellite to have the capability to obtain stereoscopic images. The platform housing the power supplies and the communications links will be adapted for ERS-1, the optical scanner on the platform being replaced by the SAR antenna. The nominal orbit is sun-synchronous and circular at an altitude of 777 km providing for a three day repetition ground track pattern.

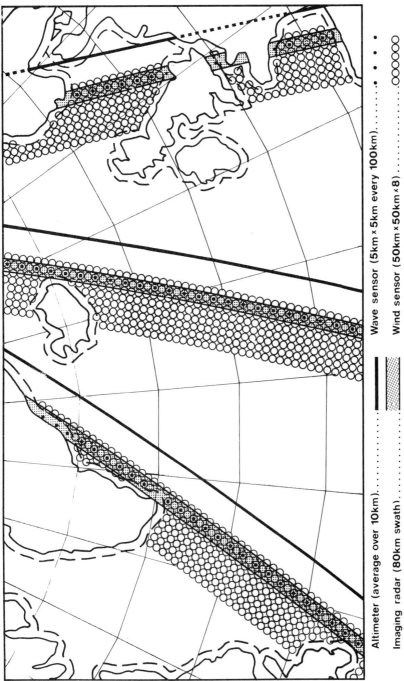

Altimeter (average over 10km)..............
Imaging radar (80km swath)..............
Wave sensor (5km × 5km every 100km).......... • • • •
Wind sensor (50km × 50km × 8)..............○○○○○

Fig. 12.4 The various 'footprints' on ERS-1 passes.

ERS-2 will be essentially similar to ERS-1 and is there as a back-up if necessary, as well as providing a continuation. ERS-3 on the other hand is a land oriented derivative of 1 and 2. This means that resources studies will have to await the launch of ERS-3 before there is a satellite designed to their requirements.

At present only one receiving station is planned, that in Kiruna in Sweden, however it is probable that other ground links will be established to give greater coverage. The data distribution system is shown diagrammatically in Fig. 12.2. Stations are, for example, proposed for Canada and possibly the USA.

The Canadian Radarsat (Figure 12.5) is primarily for ocean and ice studies, a particular requirement for Canada, and in this context it is essential to receive data in what is termed a timely fashion. It will also be used for land resource studies and could provide global data. The orbit has been especially selected to give greater coverage of the polar areas. the problem with any satellite SARs is the power requirement, thus the Radarsat will probably only be switched on for 15 to 20 minutes per orbit. This problem is difficult to overcome using conventional power from solar radiation, one solution is to use nuclear power packs but most countries are reluctant to resort to this because of environmental and political problems. The USSR does however use nuclear power for their radar satellites, very little is known about these and no imagery is available to view, it is believed that RAR rather than SAR is used hence the massive power requirement.

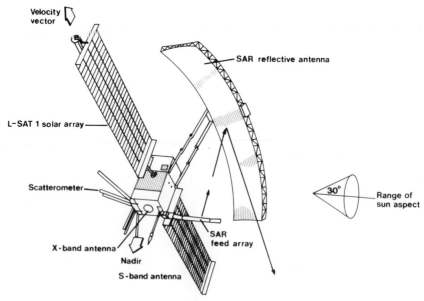

Fig. 12.5   Schematic configuration for the proposed Canadian Radarsat.

A third orbiting SAR satellite is also in advanced planning, this is to be built by the Japanese and is currently known as the J-ERS-1 and is scheduled to use L-band. Table 12.3 tabulates the essential elements of the Canadian Radarsat, the European ERS-1 and the Japanese J-ERS-1 for comparison.

Table 12.3 Comparison chart/satellite radar systems

	Radarsat	*ERS-1*	*J-ERS-1*
Launch	1990	1988	1990
Frequency	C	C	L
Processing	Par	Easier	Harder
Resolution (m)/looks	25/3	30/4	25/4
Swath width (km)	~150	80	75
Accessible swath (km)	500	80	75
Incidence angle	20°–45°	22°	33°
Sensitivity	−25 dB	−18 dB	?
Coverage	<76° N	<88° N	<80° N
Coverage interval at 73° N	<1 day	>4 days	?
Downlink	Digital compatible	>4 days	?
O/B record	Yes	No	Yes

In the long term a number of new satellites are being considered such as the FIREX (Figure 12.6) and one of the most interesting, the Ferris wheel, where thoughts return to using RAR rather than SAR and to achieve the antenna length the concept is to have a spinning chain structured like a huge wheel with a 9 km diameter. The idea is not unlike the circular casting net used by some fishermen in tropical areas, the net folds up very small but is spun around the fisherman's head in such a way that when cast it spins into a disc which falls onto the water. A chain packed into a launcher would be spun out of the cone and cast rather like the net, to form a huge spinning wheel. The system is seen as a solution to obtaining the large antenna length required for a high resolution real aperture system.

There can be little doubt that by the turn of the century great advances will have been made in the instrumentation, processing and interpretation of radar data to the extent that radar, whether airborne or space obtained, will be a major element in providing data on the earth resources and in monitoring their development and exploitation.

12.4 THE POTENTIAL IN RESOURCES STUDIES

The interpretation of a radar image does require some basic knowledge of

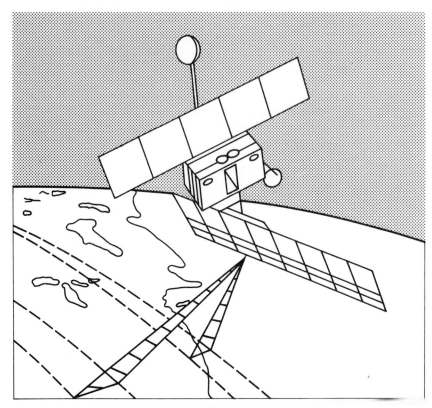

Fig. 12.6 The free flying imaging radar experiment (FIREX). Planned for polar orbiting with an L-, C- and X-band SAR as the primary instrument, producing digital data with some on-board data processing. The satellite is proposed for both land and sea observations to develop and test multiparameter data for seasonal applications and for long term studies.

the radar system and the response of targets, in many cases the criteria used to define components in optical images cannot be applied to radar. The potential use of radar in resources studies is therefore dependent to a large extent upon the ability of earth scientists and others to interpret the image. Fortunately a number of experimental campaigns have made radar imagery available to a broad spectrum of interpreters and gradually discussions on microwave technology are becoming a regular feature in technical journals.

Radar at present offers the only viable solution to the imaging of cloud covered regions and such is the urgency for knowledge of the earth's resources that this aspect alone means that it has a vital role to play. Commercial systems currently available can fulfil many of these requirements.

The early discussions made a number of references to the ability of radar images to depict the geomorphology of a region, in particular to enhance the expression of linear features. It is not surprising therefore that the major active applications of radar imagery have been in geological exploration. Couple this with the universal drive to discover sources of fossil fuels and it becomes clear why many of the surveys with airborne radars have been undertaken for geological interpretation.

In recent radar experiments such as the shuttle imaging radar SIR-A, there is evidence of surface penetration in dry desert areas, to show sub-surface patterns. In such cases the geologist has access to a data source providing information not available through any other sensing method, or at least not in such obvious clarity.

The value of the radar image to the geologist is such that it supersedes its all weather capability as a reason for acquisition. The essential data content means that radar is worth considering as an imaging system even where air photography would otherwise be readily obtainable. As a further consideration to the geologist, in many areas, data from other sources is already available and, in the search for minerals, a new system which could identify features not otherwise visible but which could improve the geological knowledge of the area must be worth exploring.

The use of radar as an imaging system is also of immediate value to the hydrologist, since it can be demonstrated that in crystalline rock areas, analysis of the fracture pattern can indicate those areas most likely to yield results from exploratory water supply drilling.

The section on penetration made reference to the potential value for snow and ice studies. Since these must be considered as valuable resources for water supply it can be shown that radar is of potential value to these studies. The all weather capability is important here if the build up of snow is to be monitored during the months when the snow is falling. In Scandinavia where the majority of power is derived from hydro-electric power stations, the availability of snow is vital to the water resources available for power and radar is already being used to provide data.

While not perhaps a resource, the ability to monitor sea ice conditions and the progress of icebergs, particularly, in areas such as the North Sea where there are numerous oil rigs, must be considered as an ancillary resource benefit. Knowledge of sea ice conditions is vital for shipping, especially in specific circumstances, including the ability to get tankers in and out of oilfields in Arctic regions or to export commodities such as timber from similar areas. Canada already uses SAR for these purposes in an operational mode and has done so for some years.

The oceans themselves can be regarded as a resource and it has already been described how much of the impetus for radar research has come from the oceanographic community. In this instance the application cannot be described as operational for there is much research still to be carried out, but it may well have a future potential.

In agriculture and forestry the application and potential have also been demonstrated. Obviously in the wet tropic regions which are under dense high forest, data on the extent and types of those forests are particularly vital. The possibility of species determination, certainly in mixed tropical dipterocarp forest would seem impossible but then nor is it at present possible by any other imaging system. A number of projects have been successfully carried out for forestry surveys, and the use of radar must be considered operational in this aspect and certainly has a future potential.

Agriculture has a requirement for imagery at specific times through the growing season and, as the growing season usually indicates a period of higher rainfall and therefore of cloud cover, radar must offer a potential imaging system alternative. In agriculture however, the requirement is to differentiate with reliability between differing crops and to assess their growth state. In this latter instance data on height, density, signs of moisture stress or infestation are all essential elements in calculating the potential yields. The radar signature can be influenced by crop row direction, leaf structure, size and angle, the presence size and shape of fruits and by motion caused by wind. These parameters make it difficult to devise a recognizable unit expression or signature for any one crop that can be extrapolated with any degree of reliability.

Recent research has shown ways in which many of these problems could be overcome, particularly with the use of multi-specular analysis and improved smoothing routines (Chapter 10).

In the future there could be an improved availability of operational multi-frequency SARs which would enable interpreters to discriminate between crops and possibly to monitor their programmed growth state. When this occurs the potential of radar for agricultural resource studies would increase.

In urban studies, although there is some evidence that major units of urban types can be differentiated, the nature of urban areas to provide bright point targets with a response beyond their target size results in urban areas being 'whited out' and internal inspection is almost impossible. Recent studies (Bryan, 1983) using squint look data provide evidence that it may be possible to extract a higher data content than was at first considered possible.

In many developing regions, one of the primary requirements is for a complete land use study of that country or region. Land use can be seen as a combination of many other disciplines to provide broad units or classifications. Thus it may only be required to define agricultural areas as orchards, arable land, irrigated and so on without specifying crops. Similarly only major tree units may be required together with the location of towns or mining works etc., and these broader units are easier to achieve. Radar has been used to good effect in a number of regional land use studies and must be considered both operational and to have further potential.

It is clear, therefore, that radar is already of proven value in resources studies, particularly in those of a broad, regional nature, but that it will be some time before it can be substituted for conventional systems in more detailed studies. The point needs to be emphasized that there is no substitute for field studies and radar is no exception to this rule. The ultimate accuracy and value of any radar interpretation is related to the level of knowledge of the interpreter, to the other data sources used and, most important, the amount of fieldwork carried out.

12.5 THE PROBLEMS

What then are the problems connected with the use of radar? Many of these have already been alluded to in the chapters dealing with the basics of systems.

The various problems connected with radar as an imaging system can be tabulated thus:

(a) SLARs are at present only available through a limited number of commercial operators, although the IRIS system is now purchasable as a system.

(b) A high degree of technical expertise particularly in electronics is essential to both acquisition and to processing.

(c) Considerable delays can occur before processed data is available because of the problems in processing data.

(d) Most radar systems are either experimental in nature and therefore in continuous change and development, or using a technology that is now out of date.

(e) The knowledge on the response of a wide variety of targets to microwaves is still very much in the research stage and much has still to be learnt about the relative effects of other parameters such as wind, moisture content, directional bias and so on.

(f) Most resources interpreters are not familiar with the radar image nor with radar systems and are often wary of using radar as a result.

(g) There is no land imaging satellite radar system comparable to the operational Landsat series or the numerous meteorological satellites.

Optical systems using cameras have the advantage that air survey cameras have reached an advanced state of development and of manufacture with the result that cameras are easily obtainable, can be readily mounted in a wide range of aircraft and do not necessarily require a sophisticated degree of processing technology. For some purposes small 'popular' cameras (usually 35 mm cameras which can be adapted to take large spools of 35 mm cine film) not originally intended for aerial photography, can be employed to good effect. The use of colour films can further extend the interpretive range of the photographic product without a great deal of technical effort. By comparison there are very few high resolution precision imagery radars which will readily fit into any aircraft, certainly not at a cost that compares with a camera. There are a number of simple radars using real time or video recording however which are lower in cost.

At the same time, apart from those real aperture radars with real time image production, all SARs require a high degree of technology in order to process the final data whether through an optical correlator or a digital processor. Figure 12.7 shows the comparative results obtained using a new high speed, dedicated digital processor that has been installed in the European Space Agency in the Netherlands.

Radar can be said to offer a wide and rewarding field for research from data acquisition and processing to interpretation.

12.6 ECONOMICS

The preceding section on problems emphasizes the fact that imaging radars, whether real or synthetic aperture, are limited in availability and are in the comparatively early stages of development.

These factors must be reflected in the economics of using radars for operational requirements. The fact that systems are not commercially produced in the way that the survey cameras are will mean that operational charges must include the high cost of the specially constructed instrumen tation. Processing has already been described as time consuming and requiring extensive computer power for digital products or a custom built optical correlator for photographic products, and once again these will be reflected in the operational costs. This problem is recognized by the manufacturers however and processing is improving with specially built fast processors being available. The cost of this equipment remains high however.

Offset against these considerations is the ability to plan a definite flying schedule with a high degree of confidence so that imagery can be obtained when planned, as against the uncertainty of obtaining optical imagery except in ideal climate conditions. This factor gains in importance when considering areas where cloud cover is persistent throughout the year.

As an example of the relative advantages that can pertain, in Nigeria where SLAR was used to obtain imagery over the whole country, a series of contracts were let to obtain aerial photographic cover nationwide in contractual blocks. In ten years cover was still not complete and a number of contractors had found the exercise uneconomic and as a result withdrew. By comparison the radar survey was planned and imagery obtained in five

Fig. 12.7 Radar images processed on the fast real time processor installed at ESA in the Netherlands.

months with very little interruptions to the flight programme. Figure 9.2 shows the status of air photography in Nigeria at the time of the radar survey, the date ranges tell their own story.

In this instance there had been considerable opposition to the survey on the grounds of data acquisition and other costs, at the conclusion of the survey it was agreed that air survey could not have been completed in the same time and that the whole survey was in fact extremely economically viable. This particular survey had been commissioned for forestry and land use and a short span time base for data was essential, even so the economic benefits were less clear than if the survey had been commissioned for mineral survey, for example, where the possible discovery of a new ore body would be more dramatic.

If the economics are between air survey under optimum conditions versus radar imagery, then there can be little doubt that the air survey will be more cost effective and that the higher resolution and geometric accuracy of the air photographs will have wider application, thereby increasing the economic benefits. Where no such benefits accrue, that is where air photography is difficult to obtain, the radar gains in economic benefit until one reaches the circumstances where radar is the only possible imaging system.

There are, however, circumstances where radar can be considered to be justified economically even though the conditions are ideally suited to photography, such circumstances are in oceanographic studies where only radar provides the wave data required, in deserts where the penetration of radar may yield additional geological data or in other areas where the particular image characteristics of radar are considered to be preferred for geological studies. The crop discrimination potential of multi-specular analysis could well become another such circumstance.

Experience with Seasat satellite radar and the shuttle radar missions indicates that in the long term the most efficient way of collecting radar data may be by the use of satellites. For one thing they provide a more stable platform and they also provide regular repeat imagery which will complement the benefits of Landsat type satellites for cloud covered areas. The experience with radar satellites and shuttle radars does, however, draw attention to the problems of data handling and to the need for the ability to handle large quantities of digital data. The provision of data processing facilities able to provide a quick turn-round of data will in all probability be reflected in high product costs.

The American Landsat satellite in ten years saw significant increases in the cost of data products, the French SPOT satellite with an improved resolution is expected to increase costs still further, radar data can be expected to be even more expensive. The use of regular image data to monitor phenomena can only be justified if image costs are economical to the user and this may mean using a different data product and not always digital data. For large areas the operational advantages of radar will therefore probably result in radar having an economic advantage over optical systems despite the seemingly initial high cost; however, the benefit has to be seen against the eventual use. For a number of purposes, such as planimetric mapping, radar would be difficult to justify until such a time as the systems improve in performance for precision mapping. Over smaller areas the use of airborne radar can show no specific economic benefit over air photography unless the particular characteristics of radar are essential to the interpretation of a particular form of data or target.

For limited applications such as monitoring oil spills in marine studies or ice flow studies, the simpler airborne SLARs producing a real time or video recorded product, using an instrumentation which can be fitted into most survey aircraft, can be considered as economically viable when set against other systems. Operational systems are in regular use for these purposes and are considered viable to avoid the costs that pollution could involve.

REFERENCE

Bryan, M. L. (1983) Urban land use classification using SAR, *Int. J. Remote Sensing,* **4,** (2) pp. 215–33.

Bibliography

1 MICROWAVE REMOTE SENSING – GENERAL PAPERS

Attema, E. P. W. (1983) The Dutch ROVE program. *IEEE Trans.* **GE 20,** (1).

Blanchard, A. J. (1982) *Measurement error considerations in polarization and phase retrieval,* IGARSS.

Blanchard, A. J., Newton, R. W., Tsang, L. and Jean, B. R. (1982) Volumetric effects and cross polarized airborne radar data. *IEEE Trans.* **GE 20,** (1), 36–41

Blanchard, A. J., Newton, R. W. and Jean, B. R. (1983) Amplitude and phase errors involved in retrieving depolarized radar cross-section measurements. *IEEE Trans.* **GE 21,** (3).

Brown, W. M. and Porcello, L. J. (1969) An introduction to synthetic aperture radar. *IEEE Spectrum,* 52–62.

Bush, T. F. and Ulaby, F. T. (1975) Fading characteristics of panchromatic radar backscatter from selected agricultural targets. *IEEE Trans.* **GE 13,** 149–57.

Bush, T. F. and Ulaby, F. T. (1976) Variability in the measurement of radar backscatter *IEEE Trans.* **AP 24,** 896–9.

Carsey, F. (1982) *Firex mission requirements, Document for renewable resources* NASA CR-169288.

Craib, K. B. (1972) *Synthetic aperture SLAR systems and their application for regional resources analysis.* Conference of Earth resource observation in information. Analysis system in remote sensing of earth resources, 1 pp. 152–78.

Daily, M. I. and Bichnell, J. J. (1981) *SAR squint analysis of directional extended targets,* IGARSS, June 8–10.

De Loor, G. P. (1980) Survey of radar application in agriculture in *Remote Sensing Application in Agriculture and Hydrology* (ed. G. Fraysse), Balkema, Holland.

De Loor, G. P. and Hoogeboom, P. (1982) Dutch ROVE program. *IEEE Trans.* **GE 20,** (1) 3–10.

Dieterle, G. and Schlude, F. (1979) *SAR systems and spacelab experiments for the European remote sensing space program.* Proc. 13 Int. Symp. on Remote Sensing. Env., Ann Arbor III, pp. 1491–8.

Elachi, C. H. (1982) *The SIR-A sensor and experiment,* IGARSS, 2, FA-6, June 1–4.

Elachi, C., Bickness, T., Jordon, R. L. and Wu, C. (1982) Spaceborne synthetic aperture imaging radars: applications, techniques and technology. *Proc. IEEE,* 70 (10) 1174–209.

Elachi, C., Goldstein, R. and Held, D. (1981) *Spaceborne bistatic synthetic aperture imaging radar,* IGARSS 2, pp. 944–50.

ESA/Marconi, *Phase A study of earth resources SAR,* Executive summary, Final report prepared by Marconi Research Laboratory.

Gjessing, D. T. and Hjelmstad, J. (1982) *Adaptive radar in remote sensing using space, frequency in polarization processing.* IEEE Radar Conf. No. 216.

Graham, L. (1976) *Earth resources radar stereo considerations,* Goodyear Aerospace Corp., Arizona Division, AEEM-550, p. 13.

Graham, L. and Rydstrom, H. O. (1974), *Synthetic aperture radar applications to earth resources development.* Goodyear Aerospace Corp. Report GERA-2010, Code 99696.

Granger, J. L. (1981) *The shuttle imaging radar,* IGARSS 2, pp. 907–12.

Guignard, J. P. (1981) *Overview of Digital Processing of SAR Data.* Proc. 15 Int. Symp. of Mem. Sens. Env., Ann Arbor, Michigan, 1, pp. 179–94.

Guignard, J., Bruzzi, S. and Pike, T. (1982) *Characterization of SAR images,* IGARSS, Munich, 2, TA-5, June 1–4.

Harger, R. O. (1970) *Synthetic Aperture Radar Systems, Theory and Design,* Academic Press, New York.

Hartl, P. H. and Sieber, A. (1981) Ongoing microwave remote sensing activities for land application in Germany, IGARSS, 2, pp. 913–19.

Haskell, A. (1981) *The European SAR 580 campaign as a preparation for ERS-1 satellite SAR.* Proc. of Int. Conf. on Matching Remote Sensing Technologies and their Applications, Remote Sensing Soc., London p. 50.

Haskell, A. and Sorensen, B. M. (1982) *The European SAR 580 Project,* IGARSS, Munich, 1, WA-5, June 1–4.

Henderson, F. M. (1975) Radar for small scale land-use mapping. *Photogram. Eng.,* XLI, (3) 307–21.

Henderson, F. M. (1977) Land use interpretation with radar imagery. *Photogram. Eng. and Remote Sensing,* 43, (1) 95–9.

Henderson, F. M. (1977) *A comparison of cell and synoptic techniques for land-use analysis with radar imagery.* Proc. Am. Soc. Photogrammetric 43 Annual Meeting.

Henderson, F. M. (1978) *Radar land-use mapping capability as a function of environmental modulation.* Proc. of Int. Symp on Remote Sensing for Observation and Inventory of Earth Resources and the Endangered Environment III, Frieburg, F. R. Germany, pp. 1547–68.

Henderson, F. M. (1979) Land use analysis of radar imagery, *Photogram. Eng. and Remote Sensing,* XLV, 295–309.

Henderson, F. M. (1979) Land use analysis for radar imagery. *Photogram. Eng. and Remote Sensing,* 45 (3).

Hughes Aircraft Company (1975) *Earth resources shuttle imaging radar*, Radar Avionics Division Aerospace Group, Culver City, California.

Inkster, D. R. and Kirby (1977) *A synthetic aperture radar (SAR) program for environmental and resource management in Canada*. Proc. 4th Canada Symposium on Remote Sensing, Quebec.

Inkster, D. R., Raney, R. K. and Dawson, R. F. (1979) *State of the art in airborne imaging radar*. Proc. 13th International Symposium on Remote Sensing, I, pp. 361–81.

Jenson, H., Graham, L. C., Porcello, L. J. and Leith, E. N. (1977) Side-looking Airborne Radar. *Science Am.,* October 1977, 84–95.

Jordon, R. L. (1980) The Seasat-A synthetic aperture radar system. *IEEE Oceanic Engineering,* **OE-5,** (2) 154–64.

Kaupp, V. H., Waite, W. P. and MacDonald, W. P. (1981) *Which angle of incidence is best for imaging radars?* IGARSS, Washington.

Kaupp, V. H., Waite, W. P. and MacDonald, H. C. (1982) Incidence angle consideration for spaceborne imaging radar. *IEEE Trans.* **GE 20.**

Koopmans, B. N. (1975) Variable flight parameters for SLAR. *Photogram. Eng. and Remote Sensing,* **XLI,** (3).

Koopmans, B. N. (1983) *Spaceborne imaging radars, present and future,* ITC Journal 1983–3.

Kovaly, J. (1976) *Synthetic Aperture Radar,* Archtec House Inc. Mass, USA.

Krul, L. (1981) *Scatterometer systems*. Proc. ESA/EARSeL Workshop, Alpback, ESA SP-166, pp. 19–27.

Krul, L. (1983) The Dutch ROVE program. *EARSeL News* (23), 13–15.

Lapgrade, G. L. (1976) *Basic concepts of synthetic aperture, side looking radar,* Goodyear Aerospace Corporation Report, Litchfield Park, Arizona.

Larson, R. W., Bagma, R. W., Ferris, J. E., Evans, M. B., Zelentza, J. S. and Soss, H. W. (1971) *Investigation of microwave hologram techniques for application to earth resources.* Proc. 5th Internat. Symp. on Remote Sensing of Environment, University of Michigan, Ann Arbor, pp. 1521–69.

Larson, R. W. and Schanda, E. (1978) *Microwave sensors and sensor systems*. Proc. 12th Int. Symp. on Remote Sensing of the Environment, Ann Arbor, University of Michigan, 1, pp. 297–305.

Leberl, F. (1979) Accuracy Analysis of Stereo SLR. *Photogram. Eng. and Remote Sensing,* **45,** (8) 1083–96.

Leberl, F., Fuchs, H. and Ford, J. P. (1981) A Radar Image Time Series. *Int J of Remote Sensing* **2,** 155–83.

Leberl, F., Jensen, H. and Kaplan, J. (1976) Side looking radar mozaicing experiments. *Photogram. Eng. and Remote Sensing* **42,** (8), 1035–43.

Leberl, F., Kobrich, M. and Domik, G. (1984) *Radiometric corrections of radar images for effects of topographic relief,* Int. Symp. on Microwave Signatures in Remote Sensing, Toulouse, France, 16–20 January.

Lewis, S. J. and Waite, W. P. (1973) Radar Shadow Frequency. *Photogram. Eng.,* **XXXIX,** (2), 189–96.

Li, F. K. and Bryan, M. L. (1983) Trade-offs among several synthetic aperture radar image quality parameters: results of a user survey. *Photogram. Eng. and Remote Sensing,* **49,** (6), 791–803.

Martin-Kaye, P. (1972) Applications of SLR in earth resource surveys in: *Environmental Remote Sensing, Applications and Achievements* (eds E. C. Barrett and L. F. Curtis).

Mégier, J., Mehl, W. and Ruppelt, R. (1984) *Methodological studies on rural land-use classification of SAR and multisensor imagery*, Proc. of the Final SAR 580 Workshop, Ispra, Italy.

Moore, R. K. (1971) Radar imagery applications: past, present and future in *Propagation Limits in Remote Sensing*, AGARD Conf. Proc. (ed. J. B. Lomax) (90).

Moore, R. K. (1979) Trade-off between picture element dimensions and non-coherent averaging in SLAR. *IEEE Trans.*, **AES-15**, (5), 697–708.

Moore, R. K. (1978) Active microwave sensing of the earth's surface. *IEEE Trans.* **AP 26** (6).

Moore, R. K., Chartant, L. J., Porcello, L. J., Stevenson, J. and Ulaby, F. T. (1975) *Microwave Remote Sensors: Handbook of Remote Sensing, American Society of Photogrammetry* pp. 399–537.

Moore, R. K., Dellwig, L. F., Parashar, S. K. and Hardy, N. E. (1974) *Application of imaging radar*, Interim Rpt. on NASA Contract NASA 7-100 Task No. RD-161, RSL Tech. Rpt. 265-2, Space Tech. Centre, Uni. Kansas, Lawrence, Kansas.

Moore, R. K. and Simonett, D. S. (1967) Potential research and earth resources studies with orbiting radars. Results of recent studies, Rpt. No. CRES-61-32, 24 pp. Pub. in Annual Meeting and Tech. Display, Anaheim, Calif., A1AA Paper 67–767, pp. 1–22.

Moore, R. K. and Waite, W. P. (1973) Octave bandwidth microwave spectral response. *Photogram. Eng.*, **38**, (10) 1051–3.

Morain, S. A. (1976) Use of radar for vegetation analysis. *Remote Sensing of the Electromagnetic Spectrum*, **3**, CL4, 61–78.

NASA (1975) *Active microwave workshop report*, NASA SP-376 (ed. R. E. Matthews).

NASA (1976) *Active Microwave workshop report*, NASA CP-2030.

Nithack, J. and Mohan, S. (1982) *The joint ISRO-DFVLR SLAR campaign 1980*, IGARSS, Munich, 2, FA-3, June 1–4.

NRCCR (1977) *Microwave remote sensing from space for earth resources surveys*, Nat. Acad. Sci., Washington, DC.

Porcello, L. J. (1977) Imaging Radar Systems. *Sci. Am.* (Oct.).

Porcello, L. J. and Rendleman, R. A. (1972) *Multispectral imaging radar* in 4th Ann. Earth Res. Program Rev 2, Rep. TM X-68397, NASA.

Raney, K. (1982) *The Canadian Radarsat Program*, IGARSS, Munich, 1, TP-6 June 1–4.

Rawson, R., Smith, F. and Larson, R. (1975) *The ERIM simultaneous X- and L-band dual polarised radar*, *IEEE* Int. Radar Conf., IEEE Pub. 75 CH0938-1, A15.

Reeves, R. G. (ed.) (1975) *Manual of Remote Sensing*, American Society of Photogrammetry, Washington, USA.

Shahin, A. and Le Toan, A. (1979) *A study of digitized radar images*, Proc. 13th Int. Symp. on Remote Sensing of the Environment, Ann Arbor, University of Michigan, 11, pp. 879–85.

Shakine, A. and Le Toan, T. (1978) *A study of digitized radar images*, Symp. of Remote Sensing of Environment, Ann Arbor, USA.

Shaw, E. (1981) *The Canadian RADAR-SAT programme,* Proc. of Int. Conf. on Matching Remote Sensing Technologies and their Applications, Remote Sensing Society, London December.

Sicco-Smit, G. (1975) Will the road to the green hell be paved with SLAR? *ITC Journal* **2**, 245–66.

Sieber, A., Freitag, B. and Lawler, K. (1982) *The aspect angle dependence of SAR images,* IGARSS, Munich, 2 FA-4, June 1–4.

Sieber, A. J., Noack, W. and Schoter, H. (1982) *Differences in two linear like polarized SAR images at the same frequency,* IGARSS, WA-5, June 10–11.

Simonett, D. S. (1970) *Remote sensing with imaging radar: a review,* Geoforum 2, pp. 61–74.

Simonett, D. S., Eagleman, J. R., Erhart, A. B., Rhodes, D. C. and Schwarz, D. E. (1967) *The potential of radar as a remote sensor in agriculture, 1, a study with K-band imagery in W. Kansas,* Univ. Kansas Centre for Research Inc., CRES Tech. Rpt. No. 61–21.

Ulaby, F. T. (1982) *Review of the approaches to the investigation of the scattering properties of material media,* IGARSS, Munich, 2 TA-1, June 1–4.

Ulaby, F. T. and Allain, C. T. (1983) Method for retrieving the true backscattering coefficient from measurements with a real antenna. *IEEE Trans.* **GE-21**, (3).

Ulaby, F. T. and Bush, T. F. (1977) *Agricultural Application of Radar,* in Proc UR51 Open internal Symp., EM Wave Propagation and Non-Ionized Media, La Baule, France, pp. 573–8.

Ulaby, F. T., Dobson, M. C., Brunfeldt, D. R. and Sieber, A. J. (1984) Calibration of the MRSE imaging radar, Int. Symp. on Microwave Signatures in Rem. Sens., Toulouse, France, January 16–20.

Ulaby, F. T., Moore, R. K. and Fung, A. K. (1981) *Microwave Remote Sensing Active and Passive,* Addison-Wesley Pub. Co. Inc., New York.

Ulaby, F. T., Moore, R. K., Mue, R. and Holtzman, J. (1972) *On microwave remote sensing,* Proc. 8th Int. Symp. on Remote Sensing of the Environment, ERIM, 2, pp. 1279–85.

Westinghouse (1974) *Spaceborne synthetic aperature radar pilot study, final report,* Contract NAS 5-2/951., Westinghouse Electric Corp., Systems Development Division, Baltimore, Maryland, USA.

Wright, R. (1982) Land-use applications of imagery radar: retrospect and prospect *Photogram. Record* **10,** 697–703.

2   GEOLOGY

Beatty, F. D., *et al.* (1965) *Geoscience potentials of side-looking radar* Raytheon Autometric Corporation, Alexandria, 10 September, Vol. 1, x + 99 pp., Vol. II 114 pp. (Prepared for NASA by US Army Corps of Engineers under Contract No. DA-44-009-AMC 1040(X)).

Beccasio, A. D. and Simons, J. H. (1965) *Regional geologic interpretation from side-looking airborne radar (SLAR),* presented at the 31st Annual Meeting, American Society of Photogrammetry, Washington, 28 March to 3 April, and in *Photogram. Eng.* **31,** (3) 507.

Blanchard, B. J. (1980) Some confusion factors in radar image interpretation in *Radar Geology: An Assessment*, JPL Publication 80-61.

Bodechtel, J., Hiller, K. and Munzer, U. (1979) *Comparison of SEASAT and LANDSAT data of Iceland for qualitative geological application*. European Space Agency SP-154.

Borden, R. C. and Williams, E. C. (1950) *Radar mapping of the Chicago-New York airway*, Technical development report 66, Technical Development and Evaluation Center, Civil Aeronautics Administration, US Dept. of Commerce, Indianapolis, April, 9 pp.

Bott, M. H., Day, A. A. and Masson-Smith, D. (1958) The Geological interpretation of gravity and magnetic surveys in Devon and Cornwall. Vol. 251 pp. 161–91.

Brown, R. D., Jr. (1966) *Geologic evaluation of radar imagery: San Andreas fault zone from Stevens Creek, Santa Clara County, to Musael Rock, San Mateo County, California*, USGS Technical Letter-NASA-45, August, 15 pp. (NTIS No. N70-38893).

Brown, W. E. and Saunders, R. S. (1978) *Radar backscatter from sand dunes*, NASA Tech. Memo. 79729. pp. 137–9.

Cameron, H. L. (1965a) *Radar as a surveying instrument in hydrology and geology*, 3rd Symp. on Remote Sensing of Environment. Proc. Univ. Michigan.

Cameron, H. L. (1965b) *Radar and geology*, Report 65-202, Air Force Cambridge Research Laboratories, Office of Aerospace Research USAF, Bedford, Mass., July, 11 pp. plus figs. (NTIS No. AD 624-887).

Cannon, P. J. (1974) Rock type discrimination using radar imagery in, *Remote Sensing of Earth Resources*, Vol. III, (Ed. F. Shahrokhi). The University of Tennessee Space Institute, Tullahoma, Tenn., xiii, 813 pp., pp. 339–52.

Clark, M. M. (1971) Comparison of SLAR images and small scale low-sun aerial photographs. *Geol. Soc. Amer. Bull.*, 82, (6), 1735–42.

Correa, A. C. (1980) *Geological mapping in the Amazon Jungle – a challenge to side looking radar*, Rept. of the Radar Geology Workshop, Snowmass, NASA JPL Publn. 80–61.

Daily, M., Elachi, C., Farr, T., Stromberg, W., Williams, S. and Schaber, G. (1978) *Applications of multispectral radar and LANDSAT imagery to geologic mapping in Death Valley*, JPL Publication 78-19, Pasadena, California.

Daily, M. I., Farr, T. and Elachi, C. (1979) Geologic interpretation from composited radar and LANDSAT imagery. *Photogram. Eng. and Remote Sensing*, 45, (8).

Daniels, D. J. (1977) The use of radar in geophysical prospecting. *Radar. 77*, IEE, London, pp. 540–6.

Dellwig, L. F. (1968) Pluses and minuses of radar in geological exploration, Earth Resources Aircraft Program Status Review, Vol. 1, Geology, Geography and Sensor Studies, NASA, MSC, Houston, September, pp. 14-1–14-25. (NTIS No. N71 16126).

Dellwig, L. F. (1969) An evaluation of multifrequency radar imagery of the Pisgah Crater Area, California, *Modern Geology*, 1, (1), 65–73. (Also: CRES Technical Report 118-6, September, 1968).

Dellwig, L. F., Kirk, J. N. and Jefferis, L. H. (1968), *The importance of radar look-direction in lineament pattern detection*, Presented at the Annual Meeting, Geological Society of America, Mexico City, 11–13 November.

Dellwig, L. F., Kirk, J. N. and Walters, R. L. (1966) The potential of low-resolution radar imagery in regional geologic studies. *J. Geophys. Res.* **71**, (20) 4495–998.

Dellwig, L. F., MacDonald, H. C. and Kirk, J. N. (1968) *The potential of radar in geological exploration,* Proc. 5th Symposium on Remote Sensing of the Environment, Report 4864-18-X, Willow Run Laboratories of the Institute of Science and Technology, The University of Michigan, Ann Arbor, April, (NTIS No. AD 676-327), pp. 747–63. Reprinted in: *Tulsa Geological Society Digest,* 1968, **36**, 26–42.

Dellwig, L. F. and Moore, R. K. (1966) The geological value of simultaneously produced like-and cross-polarized radar imagery. *J. Geophys. Research,* **71**, 3597.

Drake, B. (1972) Applications of side-looking radar to geologic investigations. Wagner, T., Tunnel-Site Selection by Remote Sensing Techniques, Report 10018-13F, Willow Run Laboratories of the Institute of Science and Technology, The University of Michigan, Ann Arbor, September, x + 91 pp., pp. 68–79.

Elachi, C. (1980) Spaceborne imaging radar: geologic and oceanographic applications. *Science,* **209**, (4461).

Elachi, C., Blom, R., Daily, M., Farr, T. and Saunders, R. S. (1980) Radar imaging of volcanic fields and sand dune fields: implications for VOIR in *Radar Geology: An Assessment,* JPL Publication 80-61.

Elder, C. H., Jeran, P. W. and Keck, D. A. (1974) *Geologic structure analysis using radar imagery of the coal mining area of Buchanan County, VA.* Report of Investigations 7869, US Dept. Interior, Bureau of Mines, Washington, DC, ii, 29 pp. (Report prepared by Pittsburgh Mining and Safety Research Center, Pittsburgh, PA).

Ellermeir, R. D., Fung, A. K. and Simonett, D. S. (1966) *Some empirical and theoretical interpretations of multiple polarization radar data,* Proc. 4th Symp. on Remote Sensing of the Environment, Report 4864-11-X, Willow Run Laboratories of the Institute of Science and Technology, The University of Michigan, Ann Arbor, April, (NTIS No. AD 638 919), pp. 657–70.

Ellermeier, R. D., Simonett, D. S. and Dellwig, L. F. (1967) *The use of multi-parameter radar imagery for the discrimination of terrain characteristics.* IEEE International Convention Record, Part 2, New York, March, pp. 127–35.

Ellison, J. H. and Williams, L. O. (1969) *Measurement on radar images of number of valleys per unit area as a discriminant of sedimentary rock types in the physiographic regions of Tennessee,* Remote Sensing Institute Technical Report 7, East Tennessee State University, Johnson City, 34 pp. (Contract: Office of Naval Research: N00014-67-A-0102-0001).

Evans, D. L. (1978) *Radar observations of a volcanic terrain: Askja Caldera, Iceland,* Publication 78-81, Jet Propulsion Laboratory, Pasadena, Calif., October 1, vii, 39 pp. (NTIS N78-33645).

Feder, A. M. (1957) *The application of radar in geologic exploration,* Bell Aircraft Corp. Report. Buffalo, NY.

Feder, A. M. (1959) Radar as a terrain-analysing device. *Geol. Soc. Amer. Bull.,* **70**, (12).

Feder, A. M. (1960) *Radar geology,* M.A. Thesis, Department of Geology, University of Buffalo, Buffalo, N.Y., February, xiv plus 80 plus Appendix.

Fischer, W. (1963) *An application of radar to geological interpretation,* 1st Symp. on Remote Sensing of Environment. Proc., Univ. Michigan, Ann Arbor.

Ford, J. P. (1980) SEASAT orbital radar imagery for geologic mapping: Tennessee-Kentucky-Virginia. *Amer. Assoc. Petrol. Geol. Bull.*

Froidevaux, C. M. (1980) Radar, an optimum remote sensing tool for detailed plate tectonic analysis and its application to hydrocarbon exploration in *Radar Geology: An Assessment*, JPL Publication 80-61.

Fridleifsson, I. *Lithology and structure of geothermal reservoir rocks in Iceland*, in Proc. 2nd UN Symposium on Development and Use of Geothermal Resources.

Gelnett, R. H. (1975) *Airborne remote sensors applied to engineering geology and civil works design investigations*, Technical Report TR-17621, Motorola Aerial Remote Sensing, Inc. Phoenix, Arizona, December, iv plus 22 pp. plus figs.

Gelnett, R. H. (1977) *The importance of look direction and depression angle in geologic applications of SLAR*, Microwave Remote Sensing Symp. Proc. NASA/ISC.

Gillerman, E. (1967) Investigation of cross-polarized radar on volcanic rocks, Technical report 61-25, CRES, Inc. University of Kansas, Lawrence, February, 11 pp. (NTIS No. N67-36566).

Gillerman, E. (1968) Major lineament and possible calderas defined by side-looking airborne radar imagery, St. François Mountains, Missouri, Technical Report 118-12, CRES, Inc. University of Kansas, Lawrence, October, 29 pp. (NTIS No. N69-32469).

Glushkov, V. M. (1972) *TOROS – side looking radar system and its application for sea ice condition study and for geologic explorations*, Pres. paper, 12th Congress, Int. Soc. Photogramm., Ottawa.

Goodyear Aerospace Corporation (1971) *Simplified description of the principles and applications of synthetic aperture terrain imaging radar*, Report GT3-9202, Arizona Division, Litchfield Park, May 26, viii plus 48 pp.

Graham, L. C. (1977) *Synthetic aperture radar geologic interpretation techniques*, Microwave Remote Sensing Symposium Proc. NASA/JSC. December.

Grant, T. A. and Cluff, L. S. (1974) Radar imagery in defining regional tectonic structure in *Annual Review of Earth and Planetary Sciences*, (ed. F. A. Donath), Annual Reviews Inc., Palo Alto, Calif., Vol. 4, pp. 123–45.

Hackman, R. J. (1966) *Geologic evaluation of radar imagery in southern Utah*, USGS Prof. Paper 527-D.

Hodler, T. W. (1977) *Remote sensing applications in hydro-geothermal exploration of the Northern Basin and Range Province*, PhD Thesis, Oregon State University, June, 235 pp. Dissertation Abstracts International, B Sciences and Engineering, **38**, (7) 3104-B (abstract only).

Inkster, R. and Kirby, M. (1977) *A synthetic aperture radar program for environmental and resource management in Canada*, Proc. 4th Can. Symp. on Remote Sensing, Quebec, pp. 469–74.

Jackson, P. L. (1980) Multichannel SAR in geologic interpretation: an appraisal in *Radar Geology: An Assessment*, JPL Publication 80-61.

Jefferis, L. H. (1969a) *Lineaments in the Grand Canyon Area, Northern Arizona – a radar analysis*, Report 118-9, CRES, Inc., The University of Kansas, Lawrence, February, ii + 16 pp. (NTIS N69 32799).

Jensen, H., Graham, L. C., Porcello, L. J. and Leither, E. N. (1977) Side-looking airborne radar. *Scientific American*, **237**, (4).

Kedar, E. Y. and Shin-Y-Hsu (1972) *Side-looking radar imagery applied in seismic risk mapping*, Proc. 8th International Symposium on Remote Sensing of

Environment, Report 195600-1-X, Willow Run Laboratories, Institute of Science and Technology, The University of Michigan, Ann Arbor, 2–6 October, pp. 1995–8.

Keppie, J. D. (1976) *Interpretation of PPI Radar Imagery of Nova Scotia.* Paper 76-3, Department of Mines, Halifax, Nova Scotia, ii, 31 pp. + figs, maps.

Kirk, J. N. (1969) *A regional study of radar lineament patterns in the Ouachita Mountains, McAlester Basin – Arkansas Valley, and Ozark Regions of Oklahoma and Arkansas.* MS Thesis, Department of Geology, University of Kansas, Lawrence, vi + 44 pp. + app.

Kirk, J. N. and Walters, R. L. (1966) Radar imagery, a new tool for the geologist. *The Compass of Sigma Gamma Epsilon*, **43**, (2), pp. 85–93.

Kenpper, D. H. and R. W. Marrs (1972) *Remote sensing aids geologic mapping,* Proc. 8th International Symposium on Remote Sensing of Environment, Report 196500-1-X, Willow Run Laboratories, Institute of Science and Technology, The University of Michigan, Ann Arbor, 2–6 October, pp. 1127–36.

Komarov, V. B. and Niavro, B. P. (1972) *Methodology of radar aerial survey with the side scanner TOROS in geological research.* [Metodika Radiolokatsionnoi Aeros'- emki Sistemoi Bokovogo Obzora TOROS Dlia Geologischeskikh Issledovanii]. *New methods of obtaining information by various remote sensors and its adaptation for solving geological problems* [Novye Metody Polucheniia Informatsii Razlichnymi Distantsionnymi Priemnikami i ee Obrabotki Dlia Resheniia Geologicheskikh Zadach], VIEMS, Moscow, (Russian).

Koopmans, B. N. (1973) Drainage analysis on radar imagery. *ITC Journal*, 464–79.

Koopmans, B. N. (1983) The potential of radar for geology, ITC special publication.

Lewis, A. J. (1971) *Geomorphic evaluation of radar imagery of southern Panama and northwestern Colombia.* PhD Dissertation, Department of Geography, University of Kansas, Lawrence, 178 pp (NTIS No. 724 118). CRES Technical Report 133-18, and US Army Topographic Command, Engineering Topographic Laboratories, Report ETL-CR-71-2.

Lewis, A. J. and MacDonald, H. C. (1971) *Radar geomorphology of Garachine Bay, Panama, and the Attrato Delta, Colombia.* Presented at the Annual Meeting, Association of American Geographers, Kansas City, March.

Lewis, A. J. and MacDonald, H. C. (1973) *Radar geomorphology of coastal and wetland environments,* Proc. American Society of Photogrammetry. Fall meeting October Part II pp. 992–1003.

Lewis, A. J. and Waite, W. P. (1972) *Relative relief measurements from radar shadows : methods and evaluation,* Proc. 68th Annual Meeting of Association of American Geographers, Kansas City, MO., 23–26 April, V. 4, pp. 65–70.

Liu, C. C. (1973a) *Geology of the area 'Senhor do Bonfim' based on SLAR mosaic interpretation,* Instituto de Pesquisas Espaciais, São José dos Campos, Brasil, 14 pp.

Liu, C. C. (1973b) *Radar geological observations on the low hilly terrain amidst Piaui, Pernambuco and Bahia State Brazil,* Instituto de Pesquisas Espaciais, São José dos Campos, Brasil, 10 pp.

Love, J. D. (1970) *Generalised geologic evaluation of side-looking radar imagery of the Teton Range and Jackson Hole, Northwestern Wyoming,* US Geological Survey Open File Report-NASA-168, February, 11 pp. + figs. (NTIS No. N72 18356).

Lyon, R. J. P. and Lee, K. (1970) Remote sensing in exploration for mineral deposits. *Economic Geology*, **63**, (7) pp. 785–800. Originally presented at Alaska Remote Sensing Conference, Anchorage, Alaska, 9 December 1969.

MacDonald, H. C. (1969a) *Remote sensing techniques in petroleum exploration*, Presented at the 18th Annual Meeting, Rocky Mountain Section, American Association of Petroleum Geologists, Albuquerque, 23–26 February. *Amer. Assoc. Petrol. Geol., Bull.* **53** (1), 216.

MacDonald, H. C. (1969b) *Geologic evaluation of radar imagery from Darien Province, Panama,* PhD Dissertation, Department of Geology, The University of Kansas, Lawrence, 141 pp. (NTIS No. AD 698 346). (Also in *Modern Geology,* **1**, (1), 1–63).

MacDonald, H. C. (1973) *Imaging radar – tool for petroleum and mineral exploration,* Presented at the 54th AAPG – 47th SEPM National Meetings, Anaheim, Calif., 14–16 May. Amer. Assoc. Petrol. Geol. Bull. **57**, (4) p. 792 (abstract only).

MacDonald, H. C. (1976) Use of radar in geology in *Geoscience Application of Imaging Radar Systems,* (ed. A. J. Lewis) *Remote Sensing of the Electromagnetic Spectrum.* Vol. 3, 93–104.

MacDonald, H. C. (1980) Historical sketch – radar geology in *Radar Geology: an Assessment,* JPL Publication 80–61.

MacDonald, H. C., Brennan, P. A. and Dellwig, L. F. (1967) Geologic evaluation by radar of NASA sedimentary test site. *IEEE Trans.* **GE 5**, (3), 72–8.

MacDonald, H. C. and Dellwig, L. F. (1969) Geologic interpretation of radar imagery from eastern Panama and northwest Colombia, presented at the 82nd Annual Meeting, Geological Society of America, Atlantic City, 10–12 November. Abstracts with Programs for 1969, Geological Society of America, Part 7, pp. 140–1 (abstract only).

MacDonald, H. C. (1969) *The influence of radar look direction on the detection of selected geological features.* Proc. 6th International Symposium on Remote Sensing of the Environment, Report 31069-2-X, Willow Run Laboratories of the Institute of Science and Technology, The University of Michigan, Ann Arbor, October, pp. 637–650.

MacDonald, H. C. and Lewis, A. J. (1969) *Applications of radar imagery in geologic and geomorphic reconnaissance of tropical environments,* presented at the 3rd Annual Meeting, South-Central Section, Geological Society of America, Lawrence, 27–29 March.
Geological Society of America, Abstracts with Programs for 1969, Part 2, pp. 18–19 (abstract only).

MacDonald, H. C. and Waite, W. P. (1970) *Optimum radar depression angles for geological analysis,* Technical Report 177-9, CRES, The University of Kansas, Lawrence, August, vi + 30 pp.
(Also in: *Modern Geology,* **2**, 1971).

MacDonald, H. C. and Waite, W. P. (1972) *Remote sensing practicality: radar geology,* Proceedings, Technical Papers, Electro-Optical Systems Design Conference, New York, 12–14 September, pp. 68–78.

MacDonald, H. C. and Waite, W. P. (1973) Imaging radars provide terrain texture and roughness parameters in semi-arid environments. *Modern Geology,* **4**, pp. 145–58.

MacDonald, H. C. and Waite, W. P. (1977) Preliminary geologic evaluation of L-band radar imagery – Arkansas test site. Final report, University of Arkansas, Fayetteville, Arkansas, November, 28 pp. (NTIS N78-33644).

MacDonald, H. C. and Wing, R. S. (1972) *Petroleum exploration with radar – eastern Panama and Northwestern Colombia,* Presented at the 57th Annual Meeting, American Association of Petroleum Geologists 17–19 April, Denver (abstract only).

Malin, M. C., Evans, D. and Elachi, C. (1978) Imaging radar observations of Askja Caldera, Iceland, *Geophys. Res. Lett.* **5**, (11) pp. 931–4.

Martin-Kaye, P. H. A. (1973) *Geology of eastern and central Nicaragua – interpretation of side-looking radar imagery,* presented at the 54th AAPG – 47th SEPM National Meetings, Anaheim, CA., 14–16 May.
Amer. Assoc. of Petrol. Geol. Bull. **57**, (4), 792, (abstract only).

Martin-Kaye, P. H. A. and Williams, A. K. (1973) Radargeologic map of eastern Nicaragua, Memoriria de la IX Conferencia Inter-Guyanas, Boletin de Geologia, Publicacion Especial No. 6, Caracas, Venezuela, pp. 600–5.

McCauley, J. R. (1972) *An evaluation of radar imagery in areas of alpine glaciation,* presented at the 6th Annual Meeting of the South-Central Section of the Geological Society of America, Manhattan, Kansas, 6–8 April.
Abstracts with Programs, Geological Society of America, **4**, (4) p. 285 (abstract only).

McCoy, R. M. (1969) Drainage network analysis with K-band radar imagery. *Geographical Review,* **59**, 493–512.

Motorola Aerial Remote Sensing, Inc. (1976) An application of side-looking airborne radar for surveying geology and natural resources potential, Motorola, Inc. Phoenix, Arizona, August, 26 pp.

Nicholson, H. E., Maxwell, W. H. and Dunne, J. A. (1977) *The SEASAT-A data processing and distribution,* Oceans 77, IEEE and MTS, Los Angeles, Vol. 9c, pp. 1–7.

Norman, J. W. (1972) Geological Applications of Side-Looking Radar in *Side-looking Radar Systems and Their Potential Applications to Earth Resource Surveys,* Vol. 3, *Potential Applications of SLR to Remote Sensing of Earth Resources,* Elliott Automation Space and Advanced Military Systems, Camberley, England. Ref. No. 5795, July, vi + 135 pp. + app., pp. 64–84 (NTIS No. N73-12401, N73-12406).

Norman, J. W. (1980) Causes of some old crustal failure zones interpreted from LANDSAT images and their significance in regional mineral exploration. *Trans. B. IMM,* **89**.

Page, L. R. (1969) Geologic analysis of the X-band radar mosaic of Massachusetts. *Second Annual Earth Resources Aircraft Program Status Review,* **1**, pp. 4-1–4-19.

Pala, D., Mussakowski, R. and Wedler, E. (1980) *SEASAT-SAR data evaluation for structural and surficial geology,* Proc. 14th Congr. Int. Soc. Photogrammetry, Hamburg.

Powers, R. E. (1972) Side-look radar provides a new tool for topographic and geological surveys. *Westinghouse Engineer,* **32**, (6), 176–81.

Prostka, H. J. (1970) Geologic interpretation of a radar mosaic of Yellowstone National Park, Inter-Agency Report NASA-179, US Geological Survey, Washington, DC (NASA-CR-121425; NTIS: N71-33185), 16 pp.

Reeves, R. G. (1968a) *Use of Radar Imagery for Structural Geologic Studies,* presented at the Annual Meeting, Northeastern Section, Geological Society of America, Washington, 15–17 February.
Geological Society of America, Special Paper 121, Abstracts for 1968, Boulder, p. 369 (abstract only).

Reeves, R. G. (1968b) *Structural geologic interpretation from radar imagery,* US Geological survey open file report – NASA-102, Washington, DC, 14 pp. + figs. (NTIS No. N69 13927).

Reeves, R. G. (1968c) *Radar Geology,* McGraw-Hill Yearbook of Science and Technology, McGraw-Hill, Inc., N.Y., pp. 322–8.

Reeves, R. G. (1969) Structural geologic interpretations from radar imagery. *Geol. Soc. Amer. Bull.,* **80,** (11), 2159–64.
(See also Reeves, R. G., 1968b.)

Reeves, R. G. (1972) *Geologic analysis of remote sensor data, Bonanze project,* 57th Annual Meeting, American Association of Petroleum Geologists, Denver, 17–19 April.
Amer. Assoc. Petrol. Geol. **56,** (3) 647 (abstract only).

Reeves, R. G. and Kover, A. N. (1966) *Radar from orbit for geologic studies,* presented at the 32nd Annual Meeting of the American Society of Photogrammetry, Washington, DC, 6–11 March.
Photogram. Eng. **32,** (5) 889 (abstract only).

Richmond, G. M. (1971) *Geologic evaluation of anomalies between like-polarized and cross-polarized K-band side-looking radar imagery of Yellowstone National Park,* US Geological Survey Interagency Report – NASA-165, Washington, 35 pp. (NTIS No. N71 33374).

Roberts, R. J. (1966) *Geological evaluation of radar imagery, North-Central Nevada,* US Geological Survey Technical Letter-NASA-49, Washington, August, 15 pp. (NTIS No N70 38894).

Rouse, J. W., Jr., MacDonald, H. C. and Waite, W. P. (1969) Geoscience applications of radar sensors.
IEEE Trans. **GE 7,** (1), 1–19.

Rydstrom, H. O. (1961) *Geological map of an area in southeastern Arizona prepared from radar photography,* Goodyear Aerospace Corp., Report AAP-13730.

Rydstrom, H. O. (1966) *Interpreting local geology from radar imagery,* Proc. 4th Symposium on Remote Sensing of the Environment, Report 4864-11-X, Willow Run Laboratories of the Institute of Science and Technology, The University of Michigan, Ann Arbor, pp. 193–201.
Geol. Soc. Amer. Bull. **78** (3), pp. 429–36 (NTIS No. AD 638 919).

Rydstrom, H. O. (1970) *Geologic exploration with high resolution radar,* GIB-9193A. Goodyear Aerospace Corporation, Arizona Division, Litchfield Park, July, vi + 48 pp.

Sabins, F. R., Jr. (1973) *Geologic interpretation of radar and space imagery of California,* Presented at the 54th AAPG – 47th SEPM National meetings, Anaheim, California, 14–16 May.
Amer. Assoc. Petrol. Geol. Bull., **57,** (4), 802, (abstract only).

Sabins, F., Blom, R. and Elachi, C. (1980) SEASAT radar image of San Andreas Fault, California.
Amer. Assoc. Petrol. Geol. Bull., **64,** 619–28.

Scanvic, J., Soubari, E. (1980) Side-looking airborne radar image interpretation and geological mapping: problems and results in *Radar Geology: An Assessment,* JPL Publication 80-61.

Schaber, G. G. (1968) *Radar and infrared in geological studies of northern Arizona,* Earth Resources Aircraft Program Status Review, Vol. 1, Geology, Geography and Sensor Studies, NASA, MSC, Houston, September, pp. 13-1–13-29 (NTIS No. N71 16126).

Schaber, G. G., Berlin, G. L. and Brown, W. E. (1976) Variations in surface roughness within Death Valley, California: geologic evaluation of 25 cm wavelength radar images. *Geol. Soc. Amer. Bull.,* **87,** (29).

Schaber, G. G. and Brown, W. E. (1972) Long-wavelength radar images of northern Arizona – a geologic evaluation, USGS Prof. Paper, 800-B, B.175.

Scheps, B. B. (1962) *The history of radar geology, Proc.* 1st Symposium on Remote Sensing of the Environment, Report 4864-1-X, Willow Run Laboratories of the Institute of Science and Technology, The University of Michigan, Ann Arbor, March, pp. 79–81 (NTIS No. AD 274 155).

Stewart, H. E., Blom, R., Abrams, M. and Daily, M. (1980) Rock type discrimination and structural analysis with LANDSAT and SEASAT data in *Radar Geology: An Assessment,* JPL Publication 80-61.

Sugiura, R. and Sabins, F. (1980) The evaluation of 3 cm wavelength radar for mapping surface deposits in the Bristol Lake/Granite Mountain area, Mojave Desert, California in *Radar Geology: An Assessment,* JPL Publication 80-61.

Taranik, J. V. and Trautwein, C. M. (1977) Integration of geological remote sensing techniques in subsurface analysis in *Subsurface Geology,* (Eds LeRoy, L. W.) Colorado School Mines, Boulder, Co.

Teleki, P. and Ramseier, R. O. (1978) *The SEASAT-A synthetic aperture radar equipment,* Proc. Int. Soc. for Photogrammetry, Commission VII, Freiburg, Germany.

Vincent, R. K. (1980) The use of radar and LANDSAT data for mineral and petroleum exploration in the Los Andes Region, Venezuela: in *Radar Geology: An Assessment,* JPL Publication 80-61.

Viskne, A., Liston, T. C. and Sapp, C. D. (1969) SLR reconnaissance of Panama. *Geophysics,* **34,** (1) pp. 54–64.
Photogram. Eng. **36,** (3), 253–69.

Waite, W. P. and MacDonald, H. C. (1972) Fracture analysis with imaging radars, Annual Meeting of the American Geophysical Union, San Francisco, December 4–7.
Trans. Amer. Geophys. Union, EOS, **53,** (11), 981.
(abstract only).

Walker, G. P. L. (1958) Geology of the Reydarfjordur area, eastern Iceland, *QJGS Lond.,* **CXIV,** Pt. 3, (455).

Walker, G. P. L. (1963) The Breiddalur central volcano, eastern Iceland. *QJGS Lond.,* **119,** Pt. 1 (473).

William, Jnr., R. S. *Satellite geological and geophysical remote sensing of Iceland,* USGS. Reston, Virginia 22092.

Williams, L. O. (1968) *Radar: an aid to geologic mapping of crystalline rocks at test site 46, North Carolina.* Technical Report. 5, East Tennessee State University Remote Sensing Institute, Johnson City, 20 pp. (Contract Office of Naval Research N0001-1-C7-A-0102-0001).

Williams, R. S., Bodvarsson, A., Fridiksson, S., Palmason, G., Rist, S., Sigtryggsson, H., Thorarinsson, S. and Thorsteinsson, I. (1973) *Satellite geological and geophysical remote sensing of Iceland – preliminary results from analysis of MSS imagery,* Symposium on Significant Results Obtained from the Earth Resources Technology Satellite-1, NASA SP-327, pp. 317–27.

Wing, R. S. (1970) *Structural analysis from radar imagery, eastern Panamanian Isthmus,* US Army Topographic Command, Corps of Engineers, Engineer Topographic Laboratories, Fort Belvoir, 156 pp.

Technical Report 133-15, CRES, The University of Kansas, Lawrence (NTIS No. AD 715 322).

PhD Dissertation, Dept. of Geology, The University of Kansas, Lawrence, 1970, 192 pp.

A portion of this paper was published under: *Structural Analysis from Radar Imagery of the Eastern Panamanian Isthmus: Part 1, Modern Geology,* Vol. 2, No. 1, February 1971, pp. 1–21.

Wing, R. S. and Dellwig, L. F. (1970a) Radar expression of Virginia Dale Precambrian ring-dike complex, Wyoming, Colorado. *Bull. Geol. Soc. Amer.,* **81,** (1), 293–8.

Wing, R. S. and Dellwig, L. F. (1970b) *Tectonic development of the eastern Panamanian Isthmus as revealed through analysis of radar imagery,* Presented at the Annual Meeting, Geological Society of America, Milwaukee 11–13 November.

Wing, R. S., Overbey, Jr., W. K. and Dellwig, L. F. (1970) Radar lineament analysis, Burning Springs Area, west Virginia – an aid in the definition of Appalachian Plateau thrusts. *Bull. Geol. Soc. Amer.,* **81,** (11), 3437–44.

Wing, R. S., MacDonald, H. C. and Dellwig, L. F. (1970) *Tectonic analysis from radar – central and eastern Panama.* Meeting of the South-Central Section of the Geological Society of America, held at College Station, Texas, 2–4 April. *Geol. Soc. Amer.,* Abstracts with Programs, **2,** (4), 305–6.

Wing, R. S. and MacDonald, H. C. (1973) Radar geology-petroleum exploration technique, eastern Panama and Northwestern Colombia. *Amer. Assoc. Petrol. Geol. Bull.,* **57,** (5), 825–40.

Wing, R. S. and Mueller, J. C. (1976) *SLAR reconnaissance, Mimika-Eilander Basin, southern trough of Irian Jaya,* Proc. of the NASA Earth Resources Survey Symposium; First Comprehensive Symposium on the Practical Application of Earth Resources Survey Data, Vol. 1-B, pp. 599–604, Technical Session Presentations. Geology – Information Systems and Services.

Wise, D. U. (1969) *Pseudo-radar topographic shadowing for detection of sub-continental sized fracture systems,* Proc. 6th International Symposium on Remote Sensing of the Environment, Report No. 31069-2-X, Willow Run Laboratories of the Institute of Science and Technology, The University of Michigan, Ann Arbor, October, pp. 603–15.

3 FORESTRY AND LAND USE

Adam, G. D. and Gentle, G. C. (1980) *Prairie surface water monitoring with SAR* Intera. pp. 60–2.

Ahern, F. J., Goodenough, D. G., Grey, A. L., Ryerson, R. A. and Vilbikaitis, R. J. (1978) Simultaneous microwave and optical wavelength observations of agricultural targets. *Canadian J. Remote Sensing* **4**, (2).

Andrianov, V. A., Armand, N. A. and Kibardina, I. N. (1976) Scattering of radio waves by an underlying surface covered with vegetation. *Radio Eng. and Electron Phys. (USA)* **21**, 12–16.

Antrop, M. (1982) *Ground truth collection for a visual interpretation of SAR* 580 imagery on the B1 site in Belgium, IGARSS, Munich, 1, WA-5, June 1–4.

Attema, E. P. W. and Van Kvilenger, J. (1974) *Short range vegetation scattering,* Proc. URSI Specialist Meeting, Berne, Switzerland.

Attema, E. P. W. (1978) The radar signature of natural surface and its application in active microwave remote sensing, in (ed. J. Lund) *Surveillance of Environmental Pollution and Resources by Electromagnetic Waves,* NATO Advanced Study Series C., pp. 227–52.

Auer, S. O. and Schutt, J. B. Remote Sensing of Vegetation and Soil using Microwave Ellipsometry, Patent – 4 052 666, Pat. Appl. 677 35 215, New York, April 1978.

Babai, P. (1974) *Measurements of complex dielectric constants of soil types in X-band,* Tech. Mem. RSC-95, Remote Sensing Centre, Texas A & M Univ. College Station, Texas.

Bahar, E. and Fitzwater, M. A. (1984) Comparison of backscatter cross sections for composite rough surfaces with different means square slope. *Int. J. Remote Sensing* **5,** (2).

Bajzak, D, (1976) *Interpretation of vegetation types on SLAR and on thermal IR imagery,* Proc. of Symposium on Remote Sensing in Forestry held during the XVI IUFRO World Congress, Oslo, pp. 87–99.

Bartholome, E. (1983) Optical SAR 580 data for land use studies, an evaluation for S. Belgium, ESA SAR 580 Invest. Prelim. Rept.

Barton, I. J. (1978) A case study comparison of microwave radiometer measurements over bare soil and vegetated surfaces. *J. Geophys. Res., ***83,** (B7).

Batlivala, P. P. and Cihlar, J. (1975) *Joint soil moisture experiment (Texas): documentation of radar backscatter and ground truth data,* R.S.L. Tech. Rept. 264-1, R.S.L. Univ. Kansas, Kansas.

Batlivala, P. P. and Ulaby, F. T. (1975a) The effect of look direction on the radar return from a row crop. R.S.L. Rept. 264-3, Univ. Kansas Centre for Res. Inc., Lawrence, Kansas.

Batlivala, P. P. and Ulaby, F. T. (1975b) *Crop identification from radar imagery of the Huntingdon County, Indiana test site,* R.S.L. Tech. Rept. 177–50, Univ. of Kansas Centre for Research.

Batlivala, P. P. and Ulaby, F. T. (1975c) Effects of roughness on the radar response to soil moisture of bare ground, R.S.L. Tech. Rept. 264–5, R.S.L. Univ. Kansas, Lawrence, Kansas.

Batlivala, P. P. and Ulaby, F. T. (1976) Radar look direction and row crops. *Photogram. Eng. and Remote Sensing* **XLII** (2), 233–8.

Batlivala, P. P. and Ulaby, F. T. (1977) *Estimation of soil moisture with radar remote sensing*, ERIM Proc. of the 11th Intern. Symp. on Remote Sensing of Env. 2, 1557–66.

Batlivala, P. P. and Ulaby, F. T. (1977) *Feasibility of monitoring soil mosture using active microwave remote sensing*, R.S.L. Tech. Rept. 264–12. The Univ. Kansas Centre for Res. Inc., Lawrence, Kansas.

Beaubion, J. (1980) *Forestry mapping using SAR* Intera. Env. Consultants Ltd., Ottawa, Canada, ASP-80-1.

Bernard, R. (1981) *A C-band radar calibration for determining surface moisture*, Centre de Recherches en Physique de l'Environment (CRPE), Rue du Gal Laclerc, 192131 Issy-les Moulineaux, France.

Blanchard, A. J. (1984) *Multifrequency and multipolarization measurement of bare vegetated fields using a ground-based pulsed radar*, Int. Symp. on Microwave Signatures in Remote Sensing, Toulouse, France.

Blyth, K. (1984) *Analysis of digital radar data from SAR-580 in relation to soil/ vegetation moisture and roughness*, Proc. of the E.S.A. SAR 580 Final Workshop, Ispra, Italy May 15–16.

Bradley, G. A. and Ulaby, F. T. (1980) *Aircraft radar response to soil moisture*, R.S.L. Tech. Rept. 460–2 (AgRISTARS SM-KO-04005), Univ. Kansas 66045.

Bradley, G. A. and Ulaby, F. T. (1981) Aircraft radar response to soil moisture, *Remote Sensing of Env.*, **II**, (6), 419–38.

Brisco, B. and Protz, R. (1978) *Evaluation of high resolution SLAR on the University of Guelph test strip*, Proc. 5th Can. Symp. on Remote Sensing, Victoria, BC.

Brisco, B. and Protz, R. (1979) *Analysis of the characteristics of soil plant systems which affect the backscattering coefficient of SAR systems*, Tech. Memo. 79-02, Department of Land Resource Science, Univ. Guelph, Ontario, Canada.

Brisco, B. and Protz, R. (1980) Corn field identification accuracy using airborne radar imagery. *Canadian J. Remote Sensing* **6**, (1).

Brisco, B., Ulaby, F. T. and Dobson, M. (1983) *Land cover and crop type classification with space SAR and Landsat data*, IGARSS, I, PS-2. August 31st–September 2nd.

Brisco, B., Ulaby, F. T. and Dobson, M. C. (1983b) *Spaceborne SAR data for land cover classification and change detection*, IGARSS, San Fran. August 31–September 2.

Brisco, B., Ulaby, F. T. and Protz, R. (1983) Improvements in crop classification accuracy with airborne radar imagery. *Photogram. Eng. and Remote Sensing*, **48.**

Brunfeldt, D. R., Ulaby, F. T. and Dobson, M. C. (1984) *Active and passive microwave observations of vegetation canopies*, Int. Symp. on Microwave Signatures in Remote Sensing, Toulouse, France. January 16–20.

Bryan, M. L. (1974) *Extraction of urban land cover data from multiplexed SAR imagery*, Proc. 9th Symp. on Remote Sensing of the Environment, Univ. Michigan, Ann Arbor, 271–89.

Bryan, M. L. (1975) Interpretation of an urban scene using multi-channel radar imagery. *Remote Sensing of Env.*, 4 (1), 49–66.

Bryan, M. L. (1979) Effect of radar azimuth angle on cultural data. *Photogram. Eng. and Remote Sensing* 45 (8), 1097–107.

Bryan, M. L. (1981) *Potentials for change detection using Seasat synthetic aperture radar data*, IGARSS, June 8–10.

Bryan, M. L. (1983) Urban land use classification using SAR. *Int. J. Remote Sensing*, 4, (2), 215–33.

Burke, H. K. and Schmugge, T. J. (1982) Effects of varying soil moisture contents and vegetation canopies on microwave emissions. *IEEE Trans.* GE 20, (3).

Bush, T. F. and Ulaby, F. T. (1975a) *Remotely sensing wheat maturation with radar*, R.S.L. Tech. Rept. 177–55, Univ. Kansas Centre for Res. Inc. Lawrence, Kansas.

Bush, T. F. and Ulaby, F. T. (1975b) *On the feasibility of monitoring cropland with radar*, Univ. of Kansas Centre for Research.

Bush, T. F. and Ulaby, F. T. (1976) *Crop classification with radar: preliminary results*, R.S.L. Tech. Rept. 330-1, Univ. Kansas Centre for Res. Inc., Lawrence, Kansas.

Bush, T. F. and Ulaby, F. T. (1977) Cropland inventories using satellite altitude imaging radar, R.S.L. Tech. Rept. 330–4, Univ. of Kansas Centre for Res. Inc., Lawrence, Kansas.

Bush, T. F. and Ulaby, F. T. (1978) Crop inventories with radar. *Canadian J. Remote Sensing*, 4, (2), 81–7.

Bush, T. F. and Ulaby, F. T. (1978) An evaluation of radar as a crop classifier. *Remote Sensing of Env.*, 7, (1), 15–36.

Bush, T. F., Ulaby, F. T., Metzler, T. and Stilles, H. (1976) *Seasonal variations of the microwave scattering properties of deciduous trees as measured in the 1–18 GHz spectral range*, R.S.L. Tech. Rept. 177–60, Univ. Kansas Centre for Research.

Calla, O. P. N., Pillai, N. S. and Bora, V. H. (1984) *Study of the scattering characteristic of sandy soil at X and QU-bands for different look angle and polarisation combination*, Int. Symp. on Microwave Signatures in Remote Sensing, Toulouse, France, 16–20 January.

Calla, O. P. N., Pillai, N. S., Kaushik, O. P. and Sivaprasad, S. (1979) *Temporal study on paddy (rice) using X-band scatterometer*, Proc. 13th Int. Symp. on Remote Sensing Env., Ann Arbor, Michigan, II, 965–7.

Cameron, H. L. (1965) *Radar as a surveying instrument in hydrology and geology*, 3 Symp. on Remote Sensing Env., Proc. Uni. Michigan.

Canoba, C. A. (1983) SIR-A and Landsat imagery interpretation of Sierra Valazco, La Rioja Province, Argentina, *I.T.C. Journal*, 1983–3.

Chang, A. T. C., Atwater, S., Salomonson, V. V., Estes, J. E. and Simonett, D. S. (1980) *L-band radar sensing of soil moisture*, NASA Rept. NASA-TM-80628.

Choudhury, B. J. (1975) *Preliminary results of SAR soil moisture experiment*, NASA TP 1404.

Choudhury, B. J., Schmugge, J. J., Newton, R. W. and Chang, A. (1979) Effect of surface roughness on the microwave emission from soils, *J. Geophysics Research*, 84.

Churchill, P. N. and Keech, M. A. (1983) *SAR investigation of Thetford Forest*, SAR 580 Investigators Preliminary Report (ESA).

Churchill, P. N. and Keech, M. A. (1984) *Multi-frequency analysis of SAR 580 imagery for woodland determinations in Thetford Forest, England,* Proc. of the ESA SAR 580 Final Workshop, Ispra, Italy.

Cihlar, J. (1973) *Ground data acquisition procedure for microwave (MAPS) measurements,* CRES Tech. Memorandum 177–42, Univ. Kansas Centre for Research Inc., Lawrence, Kansas.

Cihlar, J. and Ulaby, F. T. (1974) *Dielectric properties of soils as a function of moisture content,* Univ. Kansas Centre for Research Inc., Tech. Rept. 177–47.

Cihlar, J. F., Ulaby, F. T. and Mueller, R. (1975) *Soil moisture detection from radar imagery of the Phoenix, Arizona test site,* R.S.L. Tech. Rept. 264–4, Univ. Kansas Centre for Research Inc., Lawrence, Kansas.

Cochrane, J. T. (1983) *Savanna ecosystems in tropical South America findings from and computerised land – resource survey,* From Proc. XIV Int. Grassland, Congress, Lexington, Kentucky, USA, (eds J. A. Smith and V. W. Hays) June 15–24.

Coiner, J. C. and Dellwig, L. F. (1972) *Similarities and differences in the interpretation of air photos and SLAR imagery,* Proc. of the Tech. Program. Electro-Optical Systems Design Conference, New York, New York, September 12–14.

Crown, P. (1980) Evaluation of radar imagery for crop identification: an interpretation of radar imagery of three spring wheat sites, Intera Env. Consultants Ltd., Rept. No. ASP-80-1, Ottawa, Canada.

Daus, S. J. and Lauer, D. T. (1971a) *Testing the usefulness of SLAR imagery for evaluating forest vegetation resources,* Final Rept. Forestry R.S.L., Univ. of California, Berkeley.

Daus, S. J. and Lauer, D. T. (1971b) *SLAR imagery for evaluating wildland vegetation resources,* Proc. Am. Soc. Photog. Fall Convention.

Davis, C. F. (1973) *Forest vegetation mapping with synthetic aperture radar,* ERIM Internal Report.

De Azevedo, L. H. A. (1971) *Radar in the Amazon,* Proc. 7th Int. Symp. on Remote Sensing of the Environment, III pp. 2303–6.

De Groof, H. (1983) *Interpretability of land-use on SAR 580 data in relation with the range direction,* EARSeL/ESA Symp. on Remote Sensing for Env. Studies, Brussels, April 26–29.

DeLoor, G. P. (1974a) *The radar ground returns part III: further measurements on the radar backscatter of vegetation and soils,* Physics Lab., TNO, Rept. PHL-974-05, The Hague, The Netherlands.

De Loor, G. P. (1974b) *Measurement of radar ground returns,* Proc. of the URSI Specialist Meeting, Bern, Switzerland 23–6 September.

De Loor, G. P. (1979) *Soil moisture determination at X-band,* Proc. of the microwave remote sensing on bare soil workshop, EARSeL, Paris.

De Loor, G. P. (1980) Survey of radar applications in agriculture in *Remote Sensing Applications Agricultural and Hydrology* (ed. G. Graysse) pp. 215–32.

De Loor, G. P. and Jurriens, A. A. (1971) *The radar backscatter of vegetation,* AGARD Conference Proc. Propagation Limitations in Remote Sensing, CP-90, No. 12, 1–7.

De Loor, G. P. and Jurriens, A. A. (1972) *Radar ground returns,* Centrale Organisatie voor Toezepast – Natvurwetenschappelijh, Onderzoek ten behoeve van den Rijksverdedigning, Physisch Laboratorium, The Hague, Rijksuerdedigingsorganisatie TNO.

De Loor, G. P., Jurriens, A. A. and Gravesteijn, H. (1974) The radar backscatter from selected agricultural crops. *IEEE Trans.* **GE-12,** 70–7.

Deane, G. C. (1980) Preliminary evaluation of Seasat-A SAR data for land-use mapping. *The Geographical Jnl.* **146,** Part 3.

Dellwig, L. F. (1969) An evaluation of multi-frequency radar imagery of the Pisgah Crater area, California. *Modern Geology,* **1,** (1), 65–73.

Dellwig, L. F. and Bare, J. E. (1978 A radar investigation of the N. Louisiana salt domes. *Photogram. Eng. and Remote Sensing,* **XLIV,** (11), 1411–19.

Dellwig, L. F., Kirh, J. N. and Jeffries, L. H. (1968) *The importance of radar look direction in lineament pattern detection,* Abstracts, Geol. Soc. of Amer. Annual Meeting, Mexico City.

Dellwig, L. F. and McVauley, J. F. (1971) *Evaluation of high resolution X-band radar in the Quachita Mountains,* CRES Tech. Rept. 117–21, Univ. Kansas Centre for Res. Inc., Lawrence, Kansas.

Deutscher, M., Holzer, M. and Wellnitz-Flemming, W. (1983) *Preliminary results achieved with SAR 580 data of the Freiburg test site,* ESA SAR 580 Invest. Preliminary Report.

Dickey, F. M., King, C., Holtzman, J. C. and Moore, R. K. (1974) Moisture dependence on radar backscatter from irrigated and non-irrigated fields (at 400 MHz and 13.3 GHz). *IEEE Trans.* **GE 12** (1) 19–22.

Dobson, M. C., Eom, H. J., Ulaby, F. T. and Fung, A. K. (1984) *Active and passive microwave sensitivity to near surface soil moisture and field flooding conditions,* Int. Symp. on Microwave Signatures in Remote Sensing, Toulouse, France, January 16–20.

Dobson, M. C., Moezzi, S., Ulaby, F. T. and Roth, E. (1983) *A simulation study of scene confusion factors in sensing soil moisture from orbital radar,* RSA Tech. Rept., Uni Kansas Centre for Res. Inc., Lawrence, Kansas, 601–1.

Dobson, M. C. and Ulaby, F. T. (1981) Microwave backscatter dependence on surface roughness, soil moisture and soil texture part 3: soil tension. *IEEE Trans.,* **GE-19,** (1).

Dobson, M. C., Ulaby, F. T. and Moezzi, S. (1982) *Assessment of radar resolution requirements for soil moisture estimation from simulated satellite imagery,* RSL Tech. Rep., Uni Kansas Centre for Res. Inc., Lawrence, Kansas, 551–2.

Dobson, M. C., Ulaby, F. T., Stiles, J., Moore, R. K. and Holtzman, J. (1981) *Resolution requirements for a soil moisture imaging radar,* IGARSS, 1. June 8–10.

Drake, B. and Patton, K. H. (1980) *Land use/cover mapping from Seasat-A radar of the greater part of the Delmarua Peninsula, USA,* Proc. 14th Internal Symposium on Remote Sensing Env., San Jose, Cost Rica, 3, 1565–75.

Drake, B. and Schuchman, R. A. (1974) *Feasibility of using multiplexed SLAR imagery for water resources management and mapping vegetation communities,* Proc. 8th International Symposium on Remote Sensing of Env., Ann Arbor, Michigan, 1, 219–231.

Drake, B., Schuchman, R. A., Bryan, M. L., Larson, R. W. and Liskow, C. L. (1974)

The application of airborne imaging radars (L and X-band) to earth resources problems, Rept. NASA-CR-139385-1; ERIM-104000-1-F.

Du Fourment, H. (1983) *Interpretability of wetlands on Seasat 1 imagery in the western Polder region of Belgium,* EARSeL/ESA Symp. on Remote Sensing for Env., Studies, Brussels, April 16–29.

ESA (1979) Proceedings of the Dec. 1979 Seasat SAR Workshop Frascati, ESA, SP-154.

ESA (1979) Proceedings of the July 1979 Seasat SAR Workshop Frascati, ESA SP-154, Reprint.

ESA (1985) *Microwave Remote Sensing Applied to Vegetation,* Proc. of Dec 1984 workshops ESA SP. 227.

Eden, M. J., McGregor, D. F. M. and Morelo, J. A. (1984) Semi-quantitative classification of rain forest terrain in Columbian Arizona using radar imagery. *Int. Jl. Remote Sensing,* 5, (2), 423–31.

Elachi, C. (1981) *Shuttle imaging radar research sensor for earth resources observations,* IGARSS, Washington DC, 1, 12–127, June 8–10.

Ellermeier, R. D., Fung, A. K. and Simonett, D. S. (1966) *Some empirical and theoretical interpretation of multiple polarization radar data,* Proc. 4th Symp. on Rem Sens. of Env., Uni. Michigan, Ann Arbor, 657–70.

Ellermeier, R. D. and Simonett, D. S. (1965) *Imaging radar on spacecraft as a tool for studying the earth,* Proc. Symp. on Electromagnetic Sensing of the Earth from Satellites, Coral Galiles, pp. L1–L23.

Engman, E. T. and Jackson, T. J. (1982) *Soil moisture remote sensing applications studies of the USDA-ARS,* IGARSS, TA7.

Eppes, T. A. and Rouse, J. W. Jnr. (1974) Viewing angle effects in radar images. *Photogram. Eng.* 40, (2), 169–73.

Estes, J. E. (1982) *Preliminary analysis of shuttle imaging radar,* IGARSS, 2, FA-6, June 1–4.

Evans, R. (1984) *Synthetic aperture radar imagery for soil survey,* Proc. of the ESA SAR 580 Final Workshop, Ispra, Italy, May 15–16.

Farr, T. (1982) *Geologic interpretation of texture in Seasat and SIR-A radar images of vegetated areas,* Proc. Int. Soc. for Photogrammetric and Remote Sensing Symp., Toulouse, France, 24-VII/1, pp. 261–70.

Fenner, R. G., Pels, G. F. and Reid, S. C. (1980) *A parametric study of tillage effects on radar backscatter terrain and sea scatter workshops report,* Washington, D. C., Printed by RSL, Kansas University.

Ford, J. P. (1980) *Effects of resolution versus speckle in space-borne radar images interpretation: a geologic-user based analysis,* 7th Internal Symposium on Machine Processing of Remotely Sensed Data with Special emphasis on range forest and wetlands assessment 132–7.

Ford, J. P., Blom, R. G., Bryan, M. L., Daily, M. I. and Dixon, T. H. (1980) *Seasat view N. America, the Caribbean and W. Europe with imaging radar,* NASA Report NASA-CR-163825; JPL-PUB-80-67.

Ford, J. P., Cimino, J. B. and Elachi, C. (1983) *Space shuttle Columbia views the world with imaging radar: the SIR-A experiment,* NASA Report NAS 1.26: 169932; JPL-PUB-82-95; NASA-CR-169932.

Frost, V. S., Stiles, J. A., Holtzmann, J. C., Dellwig, L. F. and Held, D. N. (1980) *Radar image enhancement and simulations as an aid to interpretation and training,* Proc. 14th International Symposium on Remote Sensing of the Env., San Jose, Costa Rica, 2, pp. 883–96.

Gardner, J. V. (1980) *Mapping geologic structural features in tropical forest areas using SLAR: an example from Central America,* Proc. 14th International Symposium on Remote Sensing of Env., San Jose, Costa Rica, 2, pp. 1081–8.

Geiger, F. E. and Williams, D. (1972) *Dielectric constants of soils at microwave frequencies,* Goddard Space Flight Centre, Greenbelt, Maryland.

Geller, A. F., Zotova, E. N. and Naumov, M. I. (1982) Investigation of the vegetative covers on ploughed areas by SLAR. *IEEE Trans.* **GE 20** (1).

Gelnett, R. H., Dellwig, L. F. and Bare, J. E. (1978) *Increased visibility from the invisible, a comparison of radar and Landsat in tropical environments,* Proc. 12, International Symposium on Remote Sensing of Env. Manila, Rep. of the Philippines.

Gelnett, R. H. and Gardner, J. V. (1979) Use of radar for ground water exploration in Nigeria, West Africa, Proc. 13th Int. Symp. on Rem. Sens. of the Environment, Ann Arbor, Michigan.

Gombeer, R. (1983) *Crop and land-use classification study on SAR data over Belgium,* ESA SAR 580 Invest. Preliminary Report.

Goodenough, D. G. (1980) *Comparison of SAR and MSS data for monitoring of potatoes and other agricultural crops,* Intera. Env. Consultants Ltd., Report ASP-80-1, Ottawa, Canada.

Gossman, H. and Laver, H. (1984) *Mapping of young coniferous growths by means of aerial photography and radar images,* Proc. of the ESA SAR 580 Final Workshop, Ispra, Italy, May 15–16.

Guindon, B., Harris, J. W. E., Teillet, P. M., Goodenough, D. G. and Meunier, J. F. (1980) *Integ ration of MSS and SAR data of forested regions in mountainous terrain (Canada).* Proc. 14th International Symposium on Remote Sensing of Env., San Jose, Costa Rica.

Hanmugan, K. S., Ulaby, F. T., Narayanan, V. and Dobson, C. (1981) *Crop classification using multidate/multifrequency radar data.* AgRISTARS Supporing Res. Proj. Rept. SR-K2-04220, Centre for Research Inc., Univ. Kansas, Lawrence, Kansas, p. 45.

Hanson, B. C. and Moore, R. K. (1977) *Polarization and depression angle constraints in the utilisation of SLAR for identifying and mapping surface water, marsh and wetlands,* Wash., 1976 Proc. (A 77-4743, 22–43) September 28–October 1.

Haralick, R. M., Caspall, F. and Simonett, D. S. (1970) Using radar imagery for crop discrimination a statistical conditional probability study *Remote Sensing of the Environment,* 1, 131–42.

Hardy, N. E., Coiner, J. C. and Lockman, W. D. (1971) Vegetation mapping with SLAR: Yellowstone National Park in *Propagation Limits in Remote Sensing,* AGARD Conference Proc. No. 90 (ed. J. B. Lomax).

Hardy, N. E. (1981) Photographic interpretation approach to forest regrowth monitoring using SLAR – Grand County Oregon. *Int. Remote Sensing,* **2,** (2), 135–44.

Hartl, P. H., Hölzer, M., Ruppelt, T. and Wellnitz-Flemming, W. (1984), *Land use classification of radar image segments,* Proc. of the ESA SAR 580 Final Workshop, Ispra, Italy, May 15–16.

Hartl, P. H. and Sieber, A. (1981) *On going microwave remote sensing activities for application in Germany,* IGARSS, 2, pp. 913–19.

Haruto, H., Komiyama, S. and Matsuzaka, G. (1978) Cross-polarized radar backscatter from moist soil. *Remote Sensing of Env.,* **7,** (3), 211–17.

Henden, I. and Balci, M. (1983) Interpretation of SIR-A imagery of south eastern Turkey. *ITC Journal,* 1983–3.

Henderson, F. M. (1980) *Analysis of urban area land cover using Seasat SAR data,* NASA Report E82-1026; NAS 1.26: 166723; NASA-CR-166732, p. 59.

Henderson, F. M. (1983) A comparison of SAR brightness levels and urban land cover classes. *Photogram. Eng. and Remote Sensing* **49,** (11), 1585–91.

Henderson, F. M., Wharton, S. W. and Toll, T. L. (1980) *Preliminary results of mapping urban land cover with Seasat SAR 580 imagery,* American Society of Photogrammetry 46th Annual Meeting, St. Louis, MO, ASP Technical papers, A82-1002701-43), pp. 310–17, March 9–14.

Henninger, D. L. and Carney, J. H. (1983) *Shuttle imaging radar – A(SIR-A) data as a compliment to Landsat multispectral scanner* (MSS) data, IGARSS, 2, FP-5, August 31–September 2.

Hirosawa, H., Komiyama, S. and Matsuzaka, Y. (1978) *Depolarisation of radar return from rough soil surfaces,* ESA, SP 134.

Hirosawa, H., Komiyama, S. and Matsuzaka, G. (1978) Cross-polarized radar backscatter from moist soil. *Remote Sensing of Env.,* **7,** 211–17.

Hoogeboom, P. (1983) Classification of agricultural crops in radar images. *IEEE Trans.* **GE 21,** (3).

Hoekman, H. D. (1984) *Radar backscattering of forest parcels,* Paper presented at 1984 International Symposium on Microwave Signatures in Remote Sensing, Toulouse, January 16–20.

Hoekstra, P. and Delaney, A. (1974) Dielectric properties of soils at UHF and microwave frequencies. *J. Geophys. Res.,* **79,** (11).

Hoffer, R. M., Dean, M. E., Knowlton, D. J. and Latty, R. S. (1982) *Evaluation of SLAR and simulated thermatic mapper MSS data for forest cover mapping using computer aided analysis techniques (Final report),* NASA Report NAS 1.26: 167796; E83-10153; NASA-CR-167796.

Holtzman, J. (1979) *Radar studies related to the earth resources program,* Report NASA-CR-141643; CRES-TR-177-26, March 1972.

Holtzmann, J. C. (1979) *Theoretical and experimental analysis of radar backscatter from terrain,* An Annotated Bibliography, University of Kansas Centre for Research, Inc., Lawrence, Kansas.

Holtzman, J. C., Kaupp, V. H., Stiles, J. A., Frost, U. S., Komp, E. E. and Bergain, E. S. (1979) *Seasonal effects on radar imagery as predicted by PSM simulated techniques,* 10th Annual Conf. on Simulation and Modelling, Uni. Pittsburgh, Pitts., Pennsylvania.

Horne, A. I. D. and Rothnie, B. (1984) Use of optical SAR 580 data for forest and non-woodland tree survey, Proc. of the ESA SAR 580 Final Workshop, Ispra, Italy.

Hunting Geology and Geophysics Ltd., in Association with Westinghouse Electric

Corporation (1972) *Radar survey of Nicaragua*, Report to Government of Nicaragua.

Hunting Geology and Geophysics Ltd. (1981) Final report of the data content of overland Seasat-A SAR Imagery, Report to RAE Farnborough.

Hunting Geology and Geophysics Ltd. in Association with Marconi Research Centre (1982) *Radar imagery for the evaluation of non-renewable resources*, Report to client.

Hunting Geology and Geophysics Ltd. in Association with RAE Farnborough (1983a) *Evaluation of geological data content SIR-A Imagery*, Report to client.

Hunting Geology and Geophysics Ltd. (1983b) Evaluation of digitally processed SAR 580 radar imagery over part of Derbyshire test site, UK, Report to client.

Inkster, D. R., Lowry, R. T. and Thompson, M. D. (1980) *Optimum radar resolution studies for land use and forestry applications*, Proc. 14th International Symposium on Remote Sensing, San Jose, Costa Rica, 2, pp. 865–82.

Intera Environmental Consultants Ltd. (1980) *Preliminary report Seasat SAR analysis*, Report to Atmospheric Environmental Service for Contract No. 05579-00113.

Jackson, T. J. and Schmugge, T. J. (1981) Aircraft active microwave measurements for estimating soil moisture. *Photogram. Eng. and Remote Sensing*, 47 (6).

Jackson, T. J. (1980) Profile soil moisture from surface measurements. *J. Irrigation and Drainage Division*.

Jackson, T. J. and O'Neill, P. E. *Aircraft scatterometer observations of soil moisture on rangeland watersheds*.

Jackson, T. J., Schmugge, T. J. and Wang, J. R. (1982) Passive microwave sensing of soil moisture under vegetation canopies. *Water Resources Research*, 18 (4).

Karfakis, I. K. (1983) Radar imagery interpretation of northern Greece. *ITC Journal*, 1983-3.

Kasterem, H. W. J. (1977) *Measurements on the backscatter of X-band radiation of seven crops throughout the growing season*, ROVE, Report 1975.

Keydel, W. (1983) Application and experimental verification of an empirical backscattering cross section model for the earth's surface. *IEEE Trans.* **GE 20**, (1).

Kessler, R. and Reichert, P. (1983) *First results of the evaluation of the European SAR 580 data for agricultural and forestry purposes in test site D6, Freiburg*, EARSeL/ESA Symp. on Remote Sensing for Env. Studies, Brussels, 26–29 April.

Kessler, R., Reichert, P. and Losche, P. (1983) *Feasibility of different evaluation methods with X- and C-band SAR-CV-580 data in agricultural and forestry test sites*, SAR-580 Invest. Prelim. Report (ESA).

Kessler, T. and Jano, A. (1984) *Results of an evaluation of digital SAR 580 data for land-use identification*, Proc. of the ESA SAR 580 Final Workshop Ispra, Italy.

Kimes, D. S. and Kirchner, J. A. (1983) Diurnal variations of vegetation canopy structure. *Int. J. Remote Sensing*, 4, (2) 257–71.

King, G. I. (1980) *Radar discrimination of crops*, Intera Env. Consultants Ltd., Report ASP-80-1, Ottawa, Canada.

Kirdiasheve, K. P., Chukhlantsev, A. A. and Shutko, A. M. (1979) *Microwave radiation of the earth's surface in the presence of vegetation cover*, Radiotechnika i Elektronika, NASA Tech. Trans. TM 75469 24, pp. 256–264.

Knowlton, D. J. and Hoffer, R. M. (1981) *Radar imagery for forest cover mapping* Proc. of the 7th Symposium on Machine Processing of Remotely Sensed Data with special Emphasis on Range, Forest and Wetlands Assessment, *IEEE Computer Society.*

Kobayashi, T. and Hirosawa, H. (1984) *Measurement of radar backscatter from rough soil surface by any linear and circular polarizations,* Int. Symp. on Microwave Signatures in Remote Sensing, Toulouse, France, pp. 16–20.

Koolen, A. J., Koeings, F. F. R. and Bouten, W. (1979) Remote sensing of surface roughness and top soil moisture of bare tilled soil with an X-band radar. *Neth. J. Agricultural Science,* **27,** 284–96.

Koopmans, B. N., Anton, Pacheco, C., Woldai, T. and Payazs, A. (1984) *Application of multiband synthetic aperture radar imagery and colour infra-red imagery in exploration of mineralised areas of the pyrite belt, Prov. Huelva, Spain E2 test site SAR 580 campaign,* Proc. of the ESA SAR 580 Final Workshop, Ispra, Italy. May 15–16.

Krohn, M. D., Milton, N. M. and Segal, D. B. (1983) Seasat synthetic aperture radar (SAR) response to low land vegetation types in E. Maryland and Virginia. *J. Geophysics Research (USA),* **88,** (C3), 1939–52.

La Prade, C. L. and Leonardo, E. S. (1969) Evaluations from radar imagery. *Photogram. Eng.* **XXXV** (4), 366–71.

Larson, R. W., Jackson, P., Dallaire, R., Shuchman, R. and Rawson, R. (1975) *Interpretation and measurement of multi/channel microwave SAR Imagery,* Proc. 10th International Symposium on Remote Sensing Env., Ann Arbor, Michigan, 1.

Le Toan, T. (1981) *Scatterometer measurements on crop and soil surfaces,* Proc. ESA EARSeL Workshop.

Le Toan, T. (1982) *Active microwave signatures of soil and crops: significant results from three years experiment,* IGARSS, TP-2, June 1–4.

Le Toan, M., Flouzat, G., Pausader, M., Fluhr, A. and Lopes, A. (1980) *Multifrequency radar measurements of soil parameters,* Cospar, Budapest, Hungary.

Le Toan, T., Huet, M. and Lopes, A. (1981) First results of experiment using radar for wheat growth monitoring, IGARSS, 1, p. 637.

Le Toan, T. and Pausader, M. (1981) *Active microwave signatures of soil and vegetation-covered surfaces, results of measurement programmes,* Signatures Spectrales d'en Télédétection, International Society of Photogrammetry and Remote Sensing, Avignon G-II, 6, pp. 303–3123.

Le Toan, T., Pausader, M., Flouzat, G. and Fluhr, A. (1981) *Soil moisture content and microwave backscatter in the 1–9 GHz region,* IGARSS, 1, June 8–10.

Le Toan, T., Shahin, A. and Riom, J. (1980) *Application of digitalised radar images to pine forest inventory: first results,* Proc. 14th International Symposium on Remote Sensing of the Env., San Jose, Costa Rica, 2, pp. 943–4.

Lee, J. (1980) *SAR for identifying, monitoring and predicting conditions in the forest environment in British Columbia,* Final Report of the Airborne SAR Project Intera Env. Consultant Ltd., Report ASP-80-1.

Lewis, A. J. (1971) *Cumulative frequency curves of the Darien Province, Panama.* From Propagation Limits in Remote Sensing, AGARD Conference Proc., No. 90, (ed. J. B. Lomax).

Lewis, A. J. H., McDonald, H. C. and Simonett, D. S. (1969) *Detection of linear cultural features with multi-polarized radar imagery*, Proc. 6th International Symposium on Remote Sensing of Env., Univ., Michigan, Ann Arbor.

Li, F. K., Ulaby, F. T. and Easton, J. R. (1980) *Crop classification with a Landsat-radar sensor combination*, Proc. Symposium on Machine Processing of Remotely Sensed Data, Purdue Univ., West Lafayette, June 2–6.

Lichy, D. E. (1980) *Seasat: oceangraphic and terrestrial applications*, Proc. 14th International Symposium of Remote Sensing of the Env., San Jose, Costa Rica, 2, p. 1027.

MacDonald, H. C. (1969) Geologic evaluation of radar imager from Darien Province, Panama. *Modern Geology*, **1**, 1–63.

MacDonald, H. C. and Waite, W. P. (1971) Soil moisture detection with imaging radar. *Water Resources Journal*, **7**, (1), 100–9.

MacDonald, H. C. and Waite, W. P. (1973) Imaging radar provide terrain texture and roughness parameters in semi-arid environments. *Modern Geology*, **4**, 145–58.

MacDonald, H. C., Waite, W. P. and Damariche, J. C. (1980) *Use of Seasat satellite radar imagery for the detection of standing water beneath forest vegetation*, Am. Soc. Photogrammetry, Tech. Meeting, Niagara Falls.

MacDonald, H. C., Waite, W. P., Tolman, D. N. and Borengasser, M. (1978) *Long wavelength radar for geological analysis in vegetated terrain*, Proc. Am. Soc. Photogrammetry, Fall Tech. Meeting.

Mack, A. (1980) *Factors affecting the radar backscattering from an agricultural scene*, Intera Env., Consultants Ltd., Report No. ASP-80-1, Ottawa, Canada.

Marconi Research Laboratories (1980) *Evaluation of Seasat 1 SAR: final report volume 5: study of Seasat data for optimum data processing techniques*, ESA-Earthnet, Contract 3784/78-1-SW.

Mehta, N. (1983) *Crop identification with airborne scatterometers*, IGARSS, San. Fran., 1, PS-2, August 31–September 2.

Meunlek, S. and Koopmans, B. N. (1983) The shuttle imaging radar over south peninsular, Thailand. *ITC Journal*, 1983–3.

Mo T., Schmugge, T. J. and Jackson, T. J. (1984) Calculations of radar backscattering coefficient of vegetation covered soils. *Remote Sensing of the Environment*, 15, pp. 119–33.

Mohan, S. and Nithack, J. (1982) Image enhancement for determination of agricultural fields using digital SLAR data, IGARSS, 2, FA-2.

Moore, R. K., Alaa, A., Wu, L. K. and Zoughi, R. (1974) *Measurements of the sources of scatter from vegetation*, International Symposium on Microwave Signatures in Remote Sensing, Toulouse, France, 16–20 January.

Morain, S. A. (1967) *Field studies on vegetation at Horsefly Mountain, Oregon, and its relation to radar imagery*, University of Kansas CRES Tech. Rpt. 61–62.

Morain, S. A. (1971) *Recent advances in radar applications to agriculture*, Annual Earth Resources Program Review, Agriculture, Forestry and Sensor Studies NA NASA-MCS Document 03742 II.

Morain, S. A. and Campbell, J. (1974) *Radar theory applied to generalised soil mapping*, Soil Sci. Soc. Am. Proc.

Morain, S. A. and Coiner, J. (1970) *An evaluation of fine resolution radar imagery for making agricultural determination*, University of Kansas CRES Tech. Rpt. 177–7, 70 pp. 880–87.

Morain, S. A. and Coiner, J. *An evaluation of fine resolution imagery for making agricultural determinations*, NASA, N71-15001.

Morain, S. A., Holtzmann, J. and Henderson, F. M. *Radar sensing in agriculture: a socio-economic viewpoint*, IEEE EASCON Conv. Rec., Vol. 70, pp. 280–7.

Morain, S. A. and Simonett, D. S. (1966) *Vegetation Analysis with Radar Imagery*, Proc. 4 Symposium on Remote Sensing, Env., Uni, Michigan.

Morain, S. A. and Simonett, D. S. (1967) K-band radar in vegetation mapping *Photogram. Eng.*, **33** (7).

Motorola *An application of SLAR for surveying geology and natural resources potential*, Motorola Inc., Phoenix, Arizona.

Neusch, D. R. (1982) *Augmentation of landsat MSS data by Seasat SAR imagery for agricultural inventories*, AgRISTARS Rpt. IT-E2-04223, Env. Res. Inst. of Michigan, Ann Arbor, Michigan.

Newton, R. W. (1976) *Microwave sensing and its application to soil moisture detection*, Tech. Rpt. RSC-81, Remote Sensing Centre, Texas A & M University.

Newton, R. W., Lee, S. L., Rouse Jnr., J. W. and Paris, J. F. (1974) *On the feasibility of remote monitoring of soil moisture with microwave sensors*, Proc. 9th Int. Symp. on Remote Sensing of the Environment, 1, 725–38.

Nithack, J. (1982) Visual evaluation of side looking airborne radar (SLAR) imagery, NASA Rpt. ESA-TT-734, DFVLR-FB-81-11.

Nüesch, D. R. (1982a) *Augmentation of Landsat MSS data by Seasat SAR imagery for agricultural inventories*, AgRISTARS, Env. Res. Inst. of Michigan, Ann Arbor, Michigan IT-E2-04223.

Nüesche, D. R. (1982b) *Classification of SAR imagery from an agricultural region using digital textural analysis*, ISP Comm. VII, Toulouse, September 13–17.

Nunnally, N. R. (1969) Integrated landscape analysis with radar imagery. *Remote Sensing of the Environment*, **1**, (1), 1–6.

O'Sullivan, M. R. and Bell, J. W. (1976) *Synthetic aperture radar applications study*, Final Tech. Rpt. AD-A026989; TSC-PD-B482-1, RADC-TC-76-144 Tech. Service Corp., Santa Monica, Cal. May 1975–Jan. 1976.

Pala, S. (1980) *Radar imagery for assessment of coniferous forest regeneration*, Final Report of the Airborne SAR Project, Intera Env. Cons. Ltd., Report No. ASP-80-1.

Pampaloni, P. and Paloscia, S. (1984) *Microwave features of vegetation*, International Symposium on Microwave Signatures in Remote Sensing, Toulouse, France, 16–20 January.

Paquin, R. (1980) *Crop classification study using SAR*, Intera Env. Consultants Ltd., Report ASP-80-1, Ottawa, Canada.

Parashar, S., Day, D., Ryan, J., Strong, D., Worsfold, R. and King, F. (1979) *Radar discrimination of crops*, Proc. 13th Int. Symp. Env., Uni. Michigan, Ann Arbor 2, pp. 813–23, April 23–27.

Paris, J. F. (1982) Radar remote sensing of crops, IGARSS, FA4.

Paris, J. F. (1983) Radar backscattering properties of corn and soybeans of frequencies of 1.6, 4.75 and 13.3 GHz. *IEEE Trans.*, **GE 21**, (3).

Parry, J. T. (1974) X-band radar in terrain analysis under summer and winter conditions, Proc. 2, Can. Symp. on Remote Sensing Guelph, Ontario, 2, pp. 4721–86.

Parry, J. T. (1977) *Interpretation techniques for X-band SLAR,* Proc. 4th Can. Symp on Remote Sensing, Quebec.

Parry, J. T. *Interpretation techniques for X-band SLAR radar imagery of terrain and cultural landscapes in E. Quebec and Ontario,* Proceedings (A78-43303) Ontario.

Parry, D. E. and Trevett, J. W. (1979) Mapping Nigeria's vegetation from radar. *The Geographical Journal,* **145,** Pt. 2.

Parsons, A. J. and Blacknell, C. (1984) *The use of SAR 580 imagery for identification and assessment of crop condition of fields planted to wheat,* Proc. of the Final SAR 580 Workshop, Ispra, Italy.

Pedreira, A. J. (1983) Spaceborne SAR imagery interpretation, Bahia Area, Brazil. *ITC Journal,* 1983–3.

Peiyu, J., Deli, G., Guixiong, S. and Jinyu, M. (1983) *Radar backscattering coefficients of paddy fields,* IGARSS, FP5.

Protz, R. and Brisco, B. (1980) *Analysis of soil and vegetation characteristics influencing the backscatter coefficient of a SAR system.,* Intera Env. Consultants Ltd., Rpt. No. ASP-80-1, Ottawa, Canada.

RAE (1977) *Data Content of Overland Seasat-A SAR Imagery, Final Report,* February 1981.

Rao, R. G. S. and Ulaby, F. T. (1977), Optimal spatial sampling techniques for ground truth data in microwave remote sensing of soil moisture. *Rem. Sens. of Env.* **6,** (4), 289–301.

Rebillard, P. and Evans, D. (1983) Analysis of coregistered Landsat, Seasat and SIR-A images of varied terrain types. *Geophysics Research Let.* **10,** (4), 277–80.

Roessel, J. W. (1974) SLAR Mosaics for Project RADAM. *Photog. Eng.,* 50(S).

Rosenwell, C. (1970) *Detectability of water bodies by SLAR,* CRES Tech. Rpt. 177–16.

Rubec, C. (1980) *Ecological land classification using radar imagery,* Final Rpt. of the Airborne SAR Proj., Intera Env., Cons. Ltd., Rpt. ASP-80-1.

Sadowski, F. G. and Danjoy, W. A. (1980) *Some observations on the utility of remote sensors for humid tropical forests,* Proc. 14th Int. Symp. Rem. Sens. Env., San Jose, Costa Rica 3, pp. 1799–802.

Savigear, R. A. G., Cox, N. R., Hardy, J. R., Hughes, J. F., Norman, J. W. and Roberts, E. H. (1974) *Side-looking radar systems and their potential application to earth resources surveys,* Pt. 2 Earth Science applications in ESRA-CERS Scientific and Tech. Review.

Schmugge, T. J. (1978) Remote sensing of surface soil moisture, *J. Appl. Meteor.,* **17,** 1549–57.

Schmugge, T. J. (1980) *Soil moisture sensing with microwave techniques,* Proc. 14 Int. Symp. on Rem. Sens. of Env., Ann Arbor, Michigan, 1, pp. 487–506.

Schmugge, T. J. (1983) Remote sensing of soil moistures: recent advances. *J. Appl. Metrology,* **GE 21,** (3).

Schmugge, T. J., Jackson, T. J. and McKim, H. L. (1979) Survey of IN-SITU and remote sensing methods for soil moisture determination. *Satellite Hydrology.*

Schmugge, T. J., Rango, A. and Neff, R. (1976) *Satellite microwave observations of soil moisture variation,* NASA Rpt. TM-X-71036.

Schwarz, D. E. and Caspall, F. (1968) *The use of radar in the discrimination and identification of agricultural land use,* Proc. 5th Symp. on Remote Sensing of the Environment, University of Michigan, Ann Arbor.

Sekhon, R. (1981) *Application of Seasat-1 synthetic aperture radar (SAR) data to enhance and detect geological alinements and to assist Landsat landcover class mapping,* NASA Rpt. NAS 1. 26. 172893, E83-10365 NASA-CR-172893.

Shuchman, R. A., Inkster, R. and Wride, M. (1978) *Multichannel-synthetic aperture radar sensing of forest tree species,* 5th Can. Symp. on Remote Sensing, Victoria.

Shuchman, R. A. and Lowry, R. T. (1977a) *Vegetation classification with digital X-band and L-band dual polarised SAR imagery,* 4th Can. Symp. on Remote Sensing, Quebec.

Shuchman, R. and Lowry, R. T. (1977b) Vegetation classification with digital X-band and L-band dual polarised SAR imagery. *Canadian J. Remote Sensing.*

Shuchman, R. A. and Rawson, R. F. (1975) *Dual frequency and dual polarised SAR system and experiments in agricultural assessment,* IEEE Proc. Nat. Aerosp. Election Conf., Dayton, Ohio, June 10–12.

Shuchman, R. A., Rawson, R. F. and Drake, B. *A dual frequency and dual polarisation synthetic aperture radar system and experiments in agricultural assessment,* NAECON 75 Record 133.

Sicco-Smit, G. (1974) *Practical application of radar images to tropical rain forest mapping in Colombia,* Mitteilungen der Bundesforschungsanstalt für Forst-und Holzwirtschaft, No. 99, pp. 51–64.

Sicco-Smit, G. (1978a) *SLAR for forest typing in a semi-deciduous tropical region,* Proc. of the Int. symp. on Remote Sensing for Observation and Inventory of Earth Resources and the Endangered Environment, Freiburg, F.R. Germany, 111, July 2–8.

Sicco-Smit, G. (1978b) SLAR for forest type-classification in a semi-deciduous tropical region (Brazil). *ITC Journal Int. Institute for Aerial Survey and Earth Sciences, Enschede* **3,** pp. 385–401.

Sieber, A. J. (1982) *Wind influence on the backscattering coefficient from crops,* IGARSS, Munich, 2, TA-1, June 1–4.

Sieber, A. J. and Trevett, J. W. (1983) Comparison of multifrequency band radars for crop classification, *IEEE Trans.,* **GE 21,** (3).

Simonett, D. S. (1977) *Remote sensing of cultivated and natural vegetation cropland and forest,* Proceedings A77-22187, Reading, Mass.

Simonett, D. S. (1978) *Potential applications of space radar for agricultural monitoring in the wet tropics,* Proc. Int. Symp. on Remote Sensing of the Environment, Ann Arbor, Michigan 1, pp. 427–8.

Simonett, D. S. (1968a) *Radar as a sensor in agriculture,* Earth Resources Aircraft Program Status Review, NASA Manned Spacecraft Centre, Houston.

Simonett, D. S. (1968b) *The utility of radar and other remote sensors in thematic land use mapping from spacecraft,* CRES Tech. Rpt. 117–1.

Simpson, R. B. (1969) *Geographic evaluation of radar imagery in New England,* Rpt. of USGS Grant 14-08-0001-G-8, Dartmouth College Project in Rem. Sens. Hanover, New Hampshire.

Smit, M. K. (1979) Preliminary results of an investigation into the potential of applying X-band SLR images for crop-type inventory purposes. *IEEE Trans.,* **GE 17,** (4), 303–8.

Trevett, J. W. (1980) *Earth studies from Seasat imagery,* Int. Symp. Remote Sensing of the Environment, Ann Arbor, Michigan 2, pp. 805–6.

Trevett, J. W. (1983a) *Proceedings SAR-580 investigations workshop,* (ed.), JRC ISPRA, Italy, Investigation Preliminary Rpt. JRC/ESA SA/1.07, 10.8, 3.10.

Trevett, J. W. (1983b) *Multi-frequency analysis of SAR-580 for crop discrimination, Norfolk UK,* SAR 580 Invest. Prelim. Rpt. (ESA).

Trevett, J. W. (ed.) (1983c) *Investigators preliminary report: the European SAR-580 experiment,* ESA publication.

Trevett, J. W. (ed.) (1985) The European SAR 580 Final Report ESA/JRC Publication.

Trevett, J. W. and Deane, G. C. (1984) Land cover mapping from multiband SAR-580 imagery, ESA 580 Final Workshop, Ispra, Italy.

Tsang, L., Blanchard, A. J., Newton, R. W. and Kong, J. A. (1982) A simple relation between active and passive microwave remote sensing measurements of earth terrain. *IEEE Trans.,* **GE 20,** (4).

Ulaby, F. T. (1974) Radar measurements of soil moisture content. *IEEE Trans.* **AP 22,** (2), 257–65.

Ulaby, F. T. (1974) *Vegetation and soil backscatter over the 4–18 GHz region,* Proc. of the URST Specialist Meeting, Bern, Switzerland.

Ulaby, F. T. (1975) Radar response to vegetation. *IEEE Trans.* **AP 23,** 36–45.

Ulaby, F. T. (1976a) *Variations of temporal, spectral and angular radar backscattering coefficient of vegetation,* AGARD Conf. Proc. EM Propagation Characteristics of Surface Materials and Interface Aspects, 5, pp. 1–12.

Ulaby, F. T. (1976b) *Agricultural and hydrological application of radar,* Final Rept. RSL Tech. Rpt. 177–62, Univ. Kansas Centre for Res. Inc., Lawrence Kansas.

Ulaby, F. T. (1976c) Monitoring wheat growth with radar *Photogram. Eng. and Remote Sensing,* **42** (4).

Ulaby, F. T. (1981a) Microwave response of vegetation. *Adv. Space Research,* **1,** 55–70.

Ulaby, F. T. (1981b) *Active microwave sensing of soil moisture synopsis and prognosis,* Microwave Remote Sensing of Bare Soil, EARSeL Working Group IV Report, EARSeL Secretariat, Toulouse.

Ulaby, F. T. (1982a) Radar signatures of terrain: useful monitors of renewable resources. *Proc. IEEE,* **70,** (12), 1410–28.

Ulaby F. T. (1982b) *Review of the approaches to the investigation of the scattering properties of material media,* IGARSS, TA 1.

Ulaby, F. T., Aslam, A. and Dobson, M. C. (1982) Effects of vegetation cover on the radar sensitivity to soil moisture, *IEEE Trans.,* **GE 20,** (4).

Ulaby, F. T. and Bare, J. E. (1979) Look direction modulation function of the radar backscattering coefficient of agricultural fields. *Photogram. Eng. and Remote Sensing,* **45** (11), 1495–506.

Ulaby, F. T. and Batlivala, P. P. (1975) *Measurements of radar backscatter from a hybrid of sorghum,* Univ. Kansas Centre for Res. Inc. Lawrence, Kansas, RSL Tech. Rpt. 264–2.

Ulaby, F. T. and Batlivala, P. P. (1976a) Diurnal variations of radar backscattering from a vegetation canopy, *IEEE Trans.,* **AP 24,** (1).

Ulaby, F. T. and Batlivala, P. P. (1976b) Optimum radar parameters for mapping soil moisture, *IEEE Trans., GE* **14,** (2).

Ulaby, F. T. and Batlivala, P. P. (1980), Crop identification with L-band radar. *Photogram. Eng. and Remote Sensing,* **46** (1), 101–5.

Ulaby, F. T., Batlivala, P. P. and Dobson, M. C. (1978) Microwave backscatter dependence on surface roughness, soil moisture and soil texture: Part 1 – Bare Soil. *IEEE Trans., GE* **16,** (4).

Ulaby, F. T., Bradley, G. A. and Dobson, M. C. (1979) Microwave backscatter dependence on surface roughness, soil moisture and soil texture: Part 2 – Vegetation covered soils. *IEEE Trans., GE* **17,** (2).

Ulaby, F. T. and Burns, G. (1977) *The potential use of radar for crop classification and yield estimation,* Proc. Microwave Remote Sensing Symp. Houston, Texas, NASA Johnson Space Centre, Houston, Texas.

Ulaby, F. T. and Bush, T. F. (1976a) Monitoring wheat growth with radar. *Photogram. Eng. and Remote Sensing,* **42,** (4), 557–68.

Ulaby, F. T. and Bush, T. F. (1976b) Corn growth as monitored by radar. *IEEE Trans., AP* **24,** (6), 819–28.

Ulaby, F. T., Bush, T. F. and Batlivala, P. P. (1975) Radar response to vegetation 2: 8–18 GHz band. *IEEE Trans., AP* **23,** (5), 608–18.

Ulaby, F. T., Cihlar, J. and Moore, R. K. (1974) Active microwave measurement of soil moisture. *Remote Sensing of the Environment,* **3.**

Ulaby, F. T., Dobson, C., Stiles, J., Moore, R. K. and Holtzman, J. (1981) *Evaluation of the soil moisture prediction accuracy of a space radar using simulation techniques,* RSL Tech. Rpt 429-1, Univ. Kansas Centre for Res. Inc., Lawrence, Kansas 66045.

Ulaby, F. T., Dobson, M. and Bradley, G. (1980) *Radar reflectivity of bare and vegetation covered soil,* Proc. 23 Annual Conf. Committee on Spatial Res. (COSPAR), Budapest, Hungary.

Ulaby, F. T., Dobson, M. C., Brunfeldt, D. R. and Razini, M. (1982) *The effects of vegetation cover on the radar and radiometric sensitivity to soil moisture,* IGARSS, TA1.

Ulaby, F. T., Dobson, M. C., Stiles, J., Moore, R. K., Holtzman, J. (1982) A simulation study of soil moisture estimation a space SAR *Photogram. Eng. and Remote Sensing,* **48** (4), 645–60.

Ulaby, F. T., Eyton, J. R., Li, R. Y. and Burns, G. F. *Annual repeatability of multi-date radar crop classifications,* RSL Tech. Rpt., 360–1, Univ., Kansas Centre for Research Inc., Lawrence, Kansas.

Ulaby, F. T., Jedlicka, R. P. and Allen, C. T. (1984) *Measuring and modelling the dielectric and attenuation properties of vegetation,* Int. Symposium on Microwave Signature in Remote Sensing, Toulouse, France, 16–20 January.

Ulaby, F. T., Li, R. Y. and Shanmugam, K. S. (1981a) *Crop classification using airborne radar and Landsat data,* Ag RISTARS Supporting Research, SP 9K1 04043, NASA Johnson Space Centre, Houston, Texas.

Ulaby, F. T., Li, R. Y. and Shanmugam, K. S. (1981b) *Crop classification by radar,* RSL Tech. Rpt. 360-13, Univ. Kansas Centre for Research Inc.

Ulaby, F. T., Li, R. Y. and Shanmugam, K. S. (1981c) *Crop classification by radar,* IGARSS.

Ulaby, F. T. and Moore, R. K. (1973) *Radar sensing of soil moisture,* Proc 1973 INT IEEE-GAP and USNC/URSI Meeting.

Ulaby, F. T., Razini, M. and Dobson, M. C. (1983) Effects of vegetation cover on the microwave radiometric sensitivity to soil moisture. *IEEE Trans.* **GE 21,** (1).

Viksne, A. and Liston, T. C. (1970) SLR reconnaissance of Panama. *Photogram. Eng.,* **XXXVI,** (3), 253–9.

Wagner, T. W., Ott, J. S., Rudin, F. M. and Elizondo, C. (1980) *The nature of spaceborne (Seasat) radar data of developing countries. A Costa Rica case study,* Proc. 14th Int. Symp. on Rem. Sens. Env., San Jose, Costa Rica, 2, pp. 1165–75.

Waite, W. P. and MacDonald, H. C. (1971) Vegetation penetration with K-band imaging radar. *IEEE Trans.* **GE-9,** (3), 147–55.

Waite, W., MacDonald, H., Tolman, D., Barlow, C. and Borengasser, M. (1978) *Dual polarised long wavelength radar for discrimination of agricultural land use,* Proc. ASP., Albuquerque.

Wang, J. R. *et al.* (1982) Radiometric measurements over bare and vegetated fields and 5.0 GHz frequencies. *Remote Sensing of the Environment* **12.**

Wang, J. R., O'Neill, P. E., Jackson, T. J. and Engman, E. T. (1983) Multifrequency measurements of the effects of soil moisture, soil texture and surface roughness. *IEEE Trans.* **GE 21,** (1).

Wang, J. R., Schmugge, T. J. and Williams, D. (1978) *Dielectric constants of soils at microwave and frequencies,* NASA Tech. Paper TP-1238.

Woldai, T. (1983a) Landsat and SIR-A interpretation of the Kalpin Chol and Chong Korum mountains of China. *ITC Jnl.,* 1983–3, 250–2.

Woldai, T. (1983b) Lop-Nur (China) Studied from Landsat and SIR-A imagery. *ITC Jnl.* 1983-3, 253–7.

Wooding, M. G. (1983a) *Preliminary results from the analysis of SAR 580 digital radar data for the discrimination of crop types and crop condition, Cambridge UK,* SAR 580 Invest. Prelim. Rpt. (ESA).

Wooding, M. G. (1983b) *SAR 580 for the discrimination of crop types and conditions,* Remote Sensing Society Publication, Remote Sensing for Rangeland Monitoring and management.

Wooding, M. (1984) *SAR 580 radar for the discrimination of crop types and crop conditions,* Proc. of the ESA SAR 580 Final Workshop, Ispra, Italy, May 15–16.

Wu, S. T. (1981) *Analysis of results obtained from integration of Landsat multispectral scanner and Seasat synthetic aperture radar data,* Rpt. 189, NASA Earth Resources Lab., NSTL Station, Mississippi.

Wu, S. T. (1983a) *Analysis of data acquired by synthetic aperture radar and Landsat multispectral scanner over Kershaw Count, South Carolina during the summer season,* NASA Nat. Space Tech. Lab. Earth Resources Laboratory, Rpt. 213.

Wu, S. T. (1983b) *Analysis of synthetic aperture radar data acquired over a variety of land cover,* IGARSS., San Fran., 2, FP 5, August 31–September 2.

4 RADAR MAPPING

Akowetzky, W. I. (1968) On the transformation of radar coordinates into the geodetic system, *Geodezia i Aerofotosjomka* (in Russian).

Ali, A. E. (1982) *Investigation of SEASAT-A synthetic aperture radar (SAR) for topographic mapping applications,* Unpublished PhD Thesis Glasgow University.

Bair, G. L. and Carlson, G. E. (1974) Performance comparison of techniques for obtaining stereo radar images. *IEEE Trans.,* **GE 11.**

Bair, G. L. and Carlson, G. E. (1975) Height measurement with stereo radar. *Photogram. Eng. and Remote Sensing,* **XLI.**

Berlin, L. G. and Carlson, G. E. (1974) Radar Mosaics. *The Professional Geographer,* **XXIII,** (1).

Bosman, E. R., Clerici, E., Eckhart, D. and Kubik, K. (1972) Information of points from sidelooking radar images into the map system. *Bildmessung and Luftbildwesen,* **42,** (2).

Bosman, (1971) *Project Karaka – the transformation of points from sidelooking radar images into the map system,* Final Report, Part I, Netherlands Interdepartmental Working Community for the Application and Research of Remote Sensing Techniques (NIWARS), Delft.

Carlson, G. E. (1973) An improved single flight technique for radar stereo. *IEEE Trans.,* **GE 11,** (4).

Curlander, J. (1981a) *Geometric and radiometric distortion in spaceborne SAR imagery,* Invited Paper, NASA Workshop on Registration – Rectification for Terrestrial Applications. Jet Propulsion Laboratory, Pasadena, USA, November 17–19.

Curlander, J. (1981b) *Sensor to target range determination,* IPL-Interoffice Memorandum, Jet Propulsion Laboratory, Pasadena, USA, 334 7-80-056.

Dalke, G. *et al.* (1968) Regional slopes with non stereo radar. *Photogramm. Eng.,* **XXIV.**

DBA Systems (1974) *Research studies and investigations for radar control extensions,* DBA Systems Inc., PO Drawer 550, Melbourne, Florida, Defence Documentation Center Report 530784L.

Derenyi, E. E. (1974a) SLAR Geometric Test. *Photogramm., Eng.,* **XL,** 597–604.

Derenyi, E. E. (1974b) *Metric evaluation of radar and infrared imageries,* Second Canadian Symp on Remote Sensing, University of Guelph, Geulph, Ontario.

Derenyi, E. E. (1975a) *Terrain heights from SLAR imagery,* Presented at the 41st Annual Conv. Am. Soc. Photogramm. Washington, DC.

Derenyi, E. E. (1975b) Topographic accuracy of sidelooking radar imagery. *Bildmessung und Luftwesen,* 1.

Derenyi, E. E. (1975c) *Topographic accuracy of side looking radar imagery,* Proc. of ISP Commission III Symposium. F.R.G. pp. 597–604.

Derenyi, E. E. and Szabo, I. (1980) *Planimetric accuracy of synthetic aperture radar imagery,* Proc. of ISP Congress Commission III, Hamburg, FRG., B3, pp. 142–8.

Dowideit, G. (1975) *A simulation system for theoretical analysis of radar restitution and a test by adjustment,* Proc. Comm. III, Int. Soc. Photogramm., Stuttgart, W. Germany, in Deutsche Geodaetische Kommission, Reihe, B., Heft No. 214.

Dowman, I. J. and Gibson, R. (1983) *An evaluation of SAR 580 synthetic aperture radar for map revision and monitoring,* report prepared for Ordnance Survey and Department of Environment by the Department of Photogrammetry and Surveys, University College, London.

Dowman, I. J. and Gibson, R. (1984) *An evaluation of SAR 580 synthetic aperture radar for map revision and monitoring,* Proc. of the Final SAR 580 Workshop, Ispra, Italy.

Dowman, I. J. and Morris, A. H. (1982) The use of synthetic aperture radar for mapping. *Photogram. Record* **10** (60), 687–96.

Esten, R. D. (1953) *Radar relief displacement and radar parallax,* USAERDL Report 1294, Ft. Belvoir, Virginia.

Fiore, C. (1967) Sidelooking radar restitution. *Photogramm. Eng.,* **XXXIII.**

Geier, F. (1971) *Beitrag zur Geometrie des Radarbildes,* Diss, Techn. University, Graz.

Gossman, H., Lehner, M., Saurer, A. and Pfeiffer, B. (1984) *Superimposition of thermal data with radar images survey as a substitute for the digitization of topographic maps,* Research Proc. of the ESA SAR 580 Final Workshop. Ispra, Italy.

Gracie, G. (1970) *Stereo radar analysis.* US Engineer Topographic Laboratory, Ft. Belvour, Virginia. Report FTR-1339-01.

Gracie, G, Sewell, E. D. (1972) *The metric quality of stereo radar,* Proc. of the Techn. Program, Electro-Optical Systems Design Conference, New York.

Graham, L. C. (1970) *Cartographic applications of synthetic aperture radar,* Proc. Am. Soc. Photogramm. 37th Annual Meeting and Goodyear Aerospace Corp., Report GER-1626.

Graham, L. (1975a) *Flight planning for radar stereo mapping,* Proc. Am. Soc. Photogramm. 41st Meeting, Washington.

Graham, L. (1975b) *An improved orthographic radar restitutor* Presented Paper, 12th Congress, Int. Soc. Photogramm. Ottawa Canada, and Goodyear Aerospace Corp., Report GERA-1831.

Graham, L. C. (1975c) *Geometric problems in sidelooking radar imaging,* Proc. Symp Comm. III. Int. Soc. Photogramm, Stuttgart, W. Germany, in Deutsche Geodaetische Kommission, Reihe B. Heft No. 214.

Greve, C. and Cooney, W. (1974) *The digital rectification of sidelooking radar,* Proc. Am. Soc. Photogramm., Annual Convention, Washington, DC.

Hirsch, T. H. and Van Kuilenburg, J. (1976) *Preliminary tests of the SMI-SLAR mapping quality,* Netherlands Interdepartmental Working Community for the Application and Res. of Remote Sensing (NIWARS), Internal Report, 39, Delft.

Hoogeboom, P., Binnenhade, P. and Veugen, L. (1983) *An algorithm for radiometric and geometric correction of digital SLAR data,* IGARRS, WP-4, San Francisco, 1, August 31–September 2.

Hsu, A., Bruzzi, S. and Lichtenegger, J. (1984) *Evaluation of CV-580 SAR images from mapping applications,* ASAP Convention 1984, Washington, March 11–16.

Jensen, H. (1972) Mapping with coherent-radiation focused synthetic aperture sidelooking radar in 'Operational Remote Sensing: An Interactive Seminar to Evaluate Current Capabilities'. *Am. Soc. Photogramm.*

Jensen, H. (1975) *Deformations of SLAR imagery-results from actual surveys,* Proc. Symp. Comm. III, Int. Soc. Photogram, Stuttgart, W. Germany, in Deutsche Geodaetische Kommission, Riehe B, Heft No. 214.

Konecny, G. (1975) *Approach and status of geometric restitution for remote sensing imagery.* Proc. Symp. Comm. III, Int. Soc. Photogramm., Stuttgart, W. Germany, in Deutsche Geodaetische Kommission, Reihe B, Heft No. 214.

Konecny, G. and Derenyi, E. E. (1966) *Geometric consideration for mapping from scan imagery*, Proc. 4th Symp. Remote Sensing of the Environment, Ann Arbor, Michigan.

Konecny, G., Lohmann, P. and Schuhr, W. (1984) *Practical results of geometric SAR 580 image evaluation*, Proc. of the ESA SAR 580 Final Workshop, Ispra, Italy. May 15–16.

Koopmans, B. (1974) Should stereo SLAR imagery be preferred to single strip imagery for thematic mapping, *ITC Journal Part 3*.

Korneev, Iu, N. *Analytical method for photogrammetric processing of a single radar photograph*. Geodezia i Aerofotosjomka, 2.

La Prade, G. (1963) An analytical and experimental study of stereo for radar. *Photogram. Eng.*, **29**.

La Prade, G. (1970) *Subjective consideration for stereo radar*, Goodyear Aerospace Corp, GIB G149.

Leberl, F. (1970) *Metric properties of imagery produced by sidelooking airborne radar and infrared line scan systems*, Publications of the International Institute for Aerial Survey and Earth Sciences (ITC), Series A, No. 50, Delft.

Leberl, F. (1971) Vorschlaege zur Instrumentellen Entzerrung von Abbildungen mit Seitwaerts Radar (SLAR) und Infrarotabtastsystemen. *Bildmessung und Luftbildwesen*, **39**.

Leberl, F. (1972a) *Evaluation of single strips of sidelooking radar imagery*, Arch Int. Soc. Photogrammetry, Invited Paper, 12th Congress. Ottawa, Canada.

Leberl, F. (1972b) *Radargrammetria para los interpretes de imagenes*, Centro Interamericano de Fotointerpretacion, (CIAF), Bogota, Colombia.

Leberl, F. (1974) Evaluation of SLAR image quality and geometry for Proradam. *ITC Journal*, **2**, (4) Enschede.

Leberl, F. (1975a) Radargrammetric point determination PRORADAM. *Bildmessung und Luftbildwesen*, **45**, (1).

Leberl, F. (1975b) *The geometry of, and the plotting from single strips of sidelooking airborne radar imagery*, ITC Tech. Rept., No. 1.

Leberl, F. (1976) Imaging radar application to mapping and charting. *Photogrammetria*, **32**, (3), 750–100.

Leberl, F. (1977) Radar mapping applications using single images, stereo pairs, and image blocks: method and applications. *Revista Brasileira de Carto-graphia*, **20**, 16–26.

Leberl, F. (1978) Current status and perspective of active microwave imaging for geoscience application. *ITC Journal* 1978–1.

Leberl, F. (1979) *Accuracy aspects of stereo-sidelooking radar*, JPL Publication 1979-17, Jet Propulsion Laboratory Pasadena, USA.

Leberl, F. (1980) *Preliminary radargrammetric assessment of SEASAT-SAR images*, Mitteilungen der Geodaetischen Institute No. 33, Tech. University, A-8010 Graz, pp. 59–80.

Leberl, F. and Elachi, C. (1977) *Mapping with satellite sidelooking radar*, Proceedings 2nd GDTA Symposium, St. Mande, France. pp. 451–65.

Leberl, F. and Fuchs, H. (1978) *Photogrammetric differential rectification of radar images*, Pres. Paper Symp. of comm. III of the Int. Soc. Photogrammetry. Moscow and Mittl. de. Geod. Inst. No. 33, TU-Graz, 8010, Graz, Austria.

Leberl, F., Jensen, J. and Kaplan, J. (1976) Sidelooking radar mosaicking experiment. *Photogramm. Eng. and Remote Sensing.* **XLII.**

Leberl, F. W. and Raggam, J. (1982) *Satellite radargrammetry –Phase I,* Final Report to NASA, Int. Co-operation in Space, Venus Orbital Imaging Radar, 31st March.

Leberl, F., Raggam, J. and Kobrick, M. (1982) *Stereo-sidelooking radar experiments,* IGARSS, TA-2, Munich, 2, June 1–4.

Leonardo, E. (1963) Comparison of imaging geometry for radar and photographs. *Photogramm. Eng.,* **XXXIX.**

Levine, D. (1960) *Radargrammetry,* McGraw-Hill, New York.

Levine, D. (1965) *Automatic production of contour maps from radar interferometric data,* Pres. Paper Fall Tech. Meeting, Am. Soc. Photogramm, Dayton, Ohio.

Lewis, R. B. and MacDonald, H. C. (1970) Interpretative and mozaicing problems of SLAR imagery. *Rem. Sens. of the Env.,* **1,** (4), 231–6.

Ling, C., Rasmussen, L. and Campbell, W. (1978) Flight path curvature distortion in sidelooking airborne radar imagery. *Photogramm. Eng. and Remote Sensing,* **44,** (10), 1255–60.

Loelkes, G. L. (1965) *Radar mapping imagery – its enhancements and extraction for map construction,* Pres. Paper, Fall Tech. Meeting, Am. Soc. Photogram., Dayton, Ohio.

Masry, S. E., Derenyi, E. E. and Crawley, B. G. (1976) Photomaps from non-conventional imagery. *Photogramm. Eng. and Remote Sensing.* **XLII,** (4).

Moore, R. K. (1969) Heights from simultaneous radar and infrared. *Photogramm. Eng.,* **XXXV.**

Naraghhi, M., Stromberg, W. and Daily, M. (1981), *Geometric rectification of radar imagery using digital elevation models,* Image Processing Laboratory of the Jet Propulsion Laboratory, Pasadena, USA.

Peterson, R. K. (1976) *Radar correlator geometric control,* Goodyear Aerospace Report GIB 9397, Litchfield Park, Arizona, Pres. Paper, 13th Cong. Int. Soc./ Photogramm., Helsinki, Finland.

Petrie, G. (1978) *Geometric aspects and cartographic applications of sidelooking radar,* Paper presented at the 5th Annual Conference of the Remote Sensing Society, University of Durham, 21 pp.

Petrie, G., Hsu, A. and Ali, A. E. (1984) *Geometric accuracy testing of SAR 580 and SEASAT-SAR imagery,* Proc. of the ESA SAR 580 Final Workshop, Ispra, Italy.

Pfeiffer, B. and Quiel, F. (1984) *Geometric registration land use classification of SAR-580 and airborne multispectral scanner data,* Proc. of the ESA SAR 580 Final Workshop, Ispra, Italy.

Pisaruck, M. A., Kaupp, V. H., MacDonald, H. C. and Waite, W. P. (1983) *An analysis of simulated stereo radar imagery,* IGARSS, Vol 2, FA-3, San Francisco, August 31–September 2.

Raytheon, Co. (1973) *Digital rectification of sidelooking radar (DRESLAR),* Final Report, Raytheon Co., Autometric Operation, Proep. for US Army Engineer Topographic Laboratories, Fort Belvoir, Virginia, 22060, Report ETL-CR-73-18.

Rosenfield, G. H. (1968) Stereo radar techniques, *Photogram. Eng.* **34,** 586–94.

Schanda, E. (1984) *A radargammetry experiment in a mountain region*, Pra. of the ESA SAR 580 Final Workshop, Ispra, Italy.

Scheps, B. B. (1960) To measure is to know-geometric fidelity and interpretation in radar mapping. *Photogramm. Eng.*, **XXVI.**

Schlunt, R. S. and Schmid, H. P. (1982) *Real time image registration based on the Cauchy-Schwarz inequality*, IEEE. Conf.

Schreiter, J. B. (1950) *Strip projection for radar charting*, Techn. paper 130, Mapping and Charting Laboratory, Ohio State University, Ohio.

Simonett, D. S. Calculation of ground street lengths and area from radar measurements in 'The utility of radar and other remote sensors in thematic land use mapping from spacecraft', Annual Report US Geolog. Survey Interagency Report NASA 140.

Smith, H. P. (1948) *Mapping by radar – the procedures and possibilities of a new and revolutionary method of mapping and charting*, US Air Force, Randolph Field, Texas.

Thoman, G. (1969) Distance computation on radar film (ed. D. G. Simonett), '*The utility of radar and other remote sensors in thematic land use mapping from spacecraft*', Annual Report NASA-140.

Yoritomo, K. (1972) *Methods and instruments for the restitution of radar pictures*, Arch. Int. Soc. Photogram., Inv. Paper, 12th Congress, Ottawa, Canada.

Zhurkin, I. G. and Korneyev, N. (1974) Relationship between the coordinates of terrain points and the coordinates of image points in sidelooking radar systems with an antenna along the fuselage, Geodesy, Mapping, *Photogrammetry*, **16**, (3), 140–6.

5 SPECKLE AND SPECKLE REDUCTION

Alexander, L. and Kritikos, H. (1980) *An investigation of the auto-correlation function of radar images*, GE Space Division, Philadelphia.

Bennet, J. R. and McConnel, P. R. (1980) *Consideration in the design of optional multilook processor for image quality*, Proc. 3rd Seasat-SAR Workshop on SAR Image Quality, Frascati ESA SP-172, 11–12 December.

Butman, S. and Lipes, R. G. (1975) *The effects of noise and diversity on synthetic array radar imagery*, Deep Space Network Progress Report 42–29, Jet Propulsion Laboratory, Pasadena, California, pp. 46.

Crispin, J. W. and Maffett, A. L. (1971) The practical problems of radar cross-section analysis, *IEEE Trans.*, **AES 7**, 392–5.

Dainty, J. C. (1971) Detection of images immersed in speckle noise. *Opt. Acta.*, **18**, 327–39.

Dainty, J. C. (1980) An introduction to speckle, *Proc. Soc. Photo-Optical Instrumentation Eng.*, 243, Paper 243-01.

Dasarathy, B. V. and Dasarathy, H. (1981) Edge preserving filters – aid to reliable image segmentation.

ESA *Instrumentation for preprocessing of SAR data to image form*, ESA SP-1031.

Evans, D. and Farr, T. (1983) *Development of texture signatures in SAR*, IGARSS, San Francisco, 1, TA-4, August 31–September 2.

Freitag, B., Sieber, A., Guindon, B., Goodenough, D. (1983) Enhancement of high resolution SAR imagery by adaptive filters, IGARSS, San Francisco, 1, WP-4, August 31–September 2.

Frost, V., Shanmugam, K., Holtzman, J. and Stiles, J. (1982) *A statistical model for radar images of agricultural scenes,* IGARSS, Munich, 2, TA-4, June 1–4.

Frost, V. S., Stiles, J. A. and Holtzman, J. C. (1980) *Radar image processing,* Remote Sensing Lab., Uni. Kansas, Lawrence, Technical Report TR 420-1.

Frost, V. S., Stiles, J. A., Shanmugam, K. S., Holtzman, J. C. and Smith, S. A. (1981) An adaptive filter for the smoothing of noisy radar images. *Proc. of the IEEE* **69**, (1), 133–5.

Frost, V. S. A model for radar images and its application to adaptive digital filtering of multiplicative noise. *IEEE Trans.,* **PAM 4** (2).

Goodman, J. W. (1976) Some fundamental properties of speckle. *J. Opt. Soc. Am.,* **66**, 1145–50.

Guenther, B. D., Christensen, C. R. and Jain, A. (1978) *Digital processing of speckle images,* Proc. 1978 Conf. on Pattern Recognition and Image Processing, Chicago, I 11, pp. 85–90.

Hariharan, P. and Hegadus, Z. S. (1974) Reduction of speckle in coherent imagery by spatial frequency sampling, part II. *Opt. Acta.,* **21**, (9).

Kasturi, R., Krile, T. F. and Walkup, J. F. (1982) *Signal recovery from signal dependent noise,* SPIE, 359, Applications of Digital Image Processing IV, pp. 47–52.

Kato, H. and Goodman, J. W. (1975) Non-linear filtering in coherent optical systems through halftone screen processes. *Applied Optics,* **14**, (8), 1813–24.

Kondo, K., Ichiokay, Y. and Suzuki, T. (1977) Image restoration by Wiener filtering in the presence of signal dependent noise. *Appl. Opt.,* **16.**

Korwar, V. N. Degradation of picture quality by speckle in coherent mapping systems, PhD Thesis, California, Inst. of Tech. Pasadena.

Kozma, A. and Christensen, C. R. (1976) Effects of speckle on resolution. *J. Opt. Soc. Am.,* **66**, 1257–60.

Kuan, D. T., Sawchuk, A. A., Strand, T. C. and Chavel, P. (1982a) *Nonstationary 2-D recursive filter for speckle reduction,* Proc. of 1982 IEEE Int. Conf. on Acoustics, Speech and Signal Processing, Paris, France.

Kaun, D. T., Sawchuk, A. A., Strand, T. C. and Chavel, P. (1982b) *MAP Speckle Reduction Filter for Complex Amplitude Speckle Images,* Proc. 1982 IEEE Conf. on Pattern Analysis and Machine Intelligence, Las Vegas, Nevada.

Kuan, D. T., Sawchuk, A. A. and Strand, T. C. (1982c) *Adaptive restoration of images with speckle,* SPIE, 359, Application of Digital Image Processing IV, pp. 28–38.

Lee, J. S. (1980) Digital image enhancement and noise filtering by use of local statistics. *IEEE Trans.,* **PAMI 2**, (2) 165–8.

Lee, J. S. (1981a) Speckle analysis and smoothing for synthetic aperture radar images. *Computer Graphics and Image Processing,* **17**, 24–32.

Lee, J. S. (1981b) Refined filtering of image noise using local statistics. *Computer Graphics and Image Proc.* **15**, 380–289.

Lee, J. S. (1983) A simple speckle smoothing algorith for synthetic aperture radar images. *IEEE Trans.,* **SMC 13**, (1), 85–9.

Lev, A., Zucker, S. and Rosenfield, A. (1977) Iterative enhancement of noisy images. *IEEE Trans.*, **SMC 7**, (6), 435–42.

Lim, J. S. and Nawab, H. (1981) Techniques for speckle noise removal. *Optical Engineering*, **20**, 472–80.

McKechnie, T. S. (1975) *Speckle Reduction, Laser Speckle and Related Phenomena,* (ed. J. C. Dainty), New York: Springer-Verlag.

Mehl, W. (1981) *Methods of speckle filtering in SAR image quality: selected papers from workshop,* ESA, Paris, ESA SP172.

Mitchell, R. L. (1974) Models of extended targets and their coherent radar images, *Proc. IEEE*, **62**, 754–8.

Moore, R. K. (1979) Tradeoff between picture element dimensions and noncoherent averaging in side-looking airborne radar. *IEEE Trans.*, **AES 15**, (5), 697–708.

Naderi, F. and Sawchuk, A. A. (1978) Detection of low-contrast images in film grain noise. *Applied Optics*, **17**, (18), 2883–91.

Nahi, N. E. and Naraghi, M. (1975) *A general image estimation algorithm applicable to multiplicative and non-Gaussian noise,* Uni. S. California, Image Processing Institute Semi-Annual TR. No. 620, 30 September.

Oddy, C. J. and Rye, A. J. (1983) Segmentation of SAR images using a local similarity rule. *Pattern Recognition Letters 1*, 443–9.

Pierce, J. R. and Korwar, V. N. (1978) *Effect of pixel dimensions on SAR picture quality,* Proc. Synthetic Aperture Radar Technology Conf., New Mexico, State University.

Procello, L. J., Massey, N. G., Innes, R. B. and Marks, J. M. (1976) Speckle reduction in SAR. *J. Opt. Soc. Am.*, **66**, (11).

Raney, R. K. (1980a) SAR processing of partially coherent phenomena. *Int. J. Remote Sensing*, **1**, (1) 29–51.

Raney, R. K. (1980b) *SAR response to uniform random series,* Proc. 3rd Seasat-SAR Workshop on SAR Image Quality, Frascati, ESA SP-172, December 11–12.

Smith, D. J. (1984) *Quality assessment of SAR 580 products,* Proc. of the ESA SAR 580 Final Workshop, Ispra, Italy, May 15–16.

Stiles, J. A., Frost, V. S., Shanmugan, K. S. and Holtzman, J. C. (1981) *A model for simulation and processing of radar images,* IGARSS, 1, pp. 273–83.

Tomiyasu, K. (1982), *Computer modelling of speckle in a SAR image,* IGARSS, Munich, 2, TA-4, June 1–4.

Tomiyasu, K. (1983a) Computer simulation of speckle in a synthetic aperture radar image pixel. *IEEE Trans.*, **GE 21**.

Tomiyasu, K. (1983b) *Spectral analysis of complex pulse response history of a synthetic aperture radar pixel,* IGARSS, San Francisco, 2, FA-3, August 31–September 2.

Uscinski, B. J., Ouchi, K. and Thomas, J. D. *Speckle in SAR imagery,* Oxford Computer Services Ltd.

Wu, C. (1980) *A derivation of the statistical characteristic of SAR imagery data,* Proc. 3rd Seasat-SAR Workshop on SAR Image Quality, Frascati ESA SP-172, 11–12 December.

Index